A COUNTRY YEAR

Illustrations by Lauren Jarrett

Random House · New York

A COUNTRY YEAR

LIVING THE QUESTIONS

SUE HUBBELL

Library of Congress Cataloging in Publication Data
Hubbell, Sue.
Living the questions.
1. Natural history—Missouri—Addresses, essays,
lectures. 2. Natural history—Ozark Mountains—Ad-
dresses, essays, lectures. 3. Country life—Missouri—
Addresses, essays, lectures. 4. Country life—Ozark
Mountains—Addresses, essays, lectures. I. Title.
QH105.M8H83 1986   508.778'8   85-10784
ISBN 0-394-55146-X

Manufactured in the United States of America
6 8 9 7 5

Designed by Cynthia Krupat

The Wild Things helped

. . . Be patient toward all that is unsolved in your heart and try to love the questions themselves . . . Do not . . . seek the answers, which cannot be given you because you would not be able to live them. And the point is to live everything. Live the questions now. Perhaps you will . . . gradually, without noticing it, live along some distant day into the answer.

RAINER MARIA RILKE,
Letters to a Young Poet. Letter No. 4.
Translated by M. D. Herter.
© Norton, 1934, 1935.

Acknowledgments

A book of this sort is necessarily the creation of many people whose lives have touched mine and who have helped me look at the world in a special way. But there are some few to whom I should like to make my special thanks.

Linda Skrainka's painting and conversation helped me frame the questions, and Steve Skrainka would never let me forget that I am a writer even when I wanted to. Linda Verigan, Mac Johnson and Steve Cox, professional editors all, understood me better than I understood myself and without them I would not have had the courage to keep on writing. Liz Darhansoff and Bil Gilbert carried me, often protesting wildly, into print. Liddy Tebbens and Brian Hubbell drew the map that accompanies the text. And it was Brian, with his lifelong habit of understanding what seems to be ineffable, who gave me the assurance that the right words could be found. Marty Lightwood and Asher Treat read the manuscript in several stages and made valuable suggestions.

I am grateful to all of these people.

I am grateful, also, to Black Edith, who sat on the finished pages and watched fresh ones appear from the typewriter, for I suspect that it is strong magic to have a black cat sit on a manuscript.

Contents

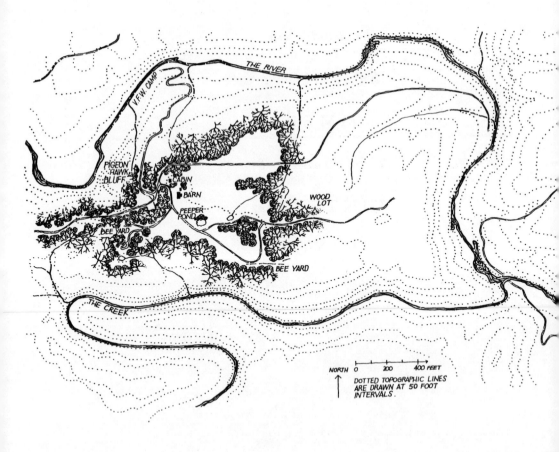

THE RIVER

V.F.W. CAMP

PIGEON
HAWK
BLUFF

CABIN

BARN

PEEPER
POND

WOOD
LOT

BEE YARD

BEE YARD

THE CREEK

NORTH

0   200   400 FEET

DOTTED TOPOGRAPHIC LINES
ARE DRAWN AT 50 FOOT
INTERVALS.

There are three big windows that go from floor to ceiling on the south side of my cabin. I like to sit in the brown leather chair in the twilight of winter evenings and watch birds at the feeder that stretches across them. The windows were a gift from my husband before he left the last time. He had come and gone before, and we were not sure that this would be the last time, although I suspected that it was.

I have lived here in the Ozark Mountains of southern Missouri for twelve years now, and for most of that time I have been alone. I have learned to run a business that we started together, a commercial beekeeping and honey-producing operation, a shaky, marginal sort of affair that never quite leaves me free of money worries but which allows me to live in these hills that I love.

My share of the Ozarks is unusual and striking. My farm lies two hundred and fifty feet above a swift, showy river to the north and a small creek to the south, its run broken by waterfalls. Creek and river join just to the east, so I live on a peninsula of land. The back fifty acres are covered with second-growth timber, and I take my firewood there. Last summer when I was cutting firewood, I came across a magnificent black walnut, tall and straight, with no jutting branches to mar its value as a timber tree. I don't expect to sell it, although even a single walnut so straight and unblemished would fetch a good price, but I cut some trees near it to give it room. The botanic name for black walnut is *Juglans nigra*—"Black Nut Tree of God," a suitable name for a tree of such dignity, and I wanted to give it space.

Over the past twelve years I have learned that a tree needs space to grow, that coyotes sing down by the creek in January, that I can drive a nail into oak only when it is green, that bees know more about making honey than I do, that love can become sadness, and that there are more questions than answers.

SPRING

The river to the north of my place is claimed by the U.S. Park Service, and the creek to the south is under the protection of the Missouri State Conservation Department, so I am surrounded by government land. The deed to the property says my farm is a hundred and five acres, but it is probably something more like ninety. The land hasn't been surveyed since the mid-1800s and it is hard to know where the boundaries are; a park ranger told me he suspected that the nineteenth-century surveyor had run his lines from a tavern, because the corners seem to have been established by someone in his cups.

The place is so beautiful that it nearly brought tears to my eyes the first time I saw it twelve years ago; I feel the same way today, so I have never much cared about the number of acres, or where the boundary lines run or who, exactly, owns what. But the things that make it so beautiful and desirable to me have also convinced others that this is prime land, too, and belongs to them as well. At the moment, for instance, I am feeling a bit of an outsider, having discovered that I live in the middle of an indigo bunting ghetto. As ghettos go, it is a cheerful one in which to live, but it has forced me to think about property rights.

Indigo buntings are small but emphatic birds. They believe that they own the place, and it is hard to ignore their claim. The male birds—brilliant, shimmering blue—perch on the garden posts or on top of the cedar trees that have taken over the pasture. From there they survey their holdings and belt out their songs, complicated tangles of couplets that waken me first thing in the morning; they keep it up all day, even at noon, after the other birds have quieted. The indigo buntings have several important facts to tell us, especially about who's in charge around here. The dull brown, sparrowlike females and juveniles are more interested in eating; they stay nearer the ground and search the low-growing shrubs and grasses for seeds and an occasional caterpillar, but even they know what's what. One day, walking back along the edge of the field, I came upon a young indigo bunting preoc-

cupied with song practice. He had not yet dared take as visible a perch as his father would have chosen, but there he was, clinging to a bare twig and softly running through his couplets, getting them all wrong and then going back over them so quietly that had I not been within a few feet of him I would not have heard.

Another time I discovered that the back door of the honey house had blown open and the room was filled with a variety of winged creatures. Most were insects, but among them I found a half-grown indigo bunting who had blundered in and was trying to find his way out, beating his small wings against the screened window. Holding him carefully, I stroked the back of his neck to try to soothe him, but discovered that his heart was not beating in terror. Perhaps he was so young that he had not learned fear, but I prefer to think that like the rest of his breed he was simply too pert and too sure of his rights to be afraid. He eyed me crossly and tweaked my giant thumb with his beak to tell me that I was to let him go right this minute. I did so, of course, and watched him fly off to the tall grasses behind the honey house, where I knew that one family of indigo buntings had been nesting.

Well, they think they own the place, and their assurance is only countered by a scrap of paper in my files. But there are other contenders, and perhaps I ought to try to take a census and judge claims before I grant them title. There are other birds who call this place theirs—buzzards, who work the updrafts over the river and creek, goldfinches, wild turkey, phoebes and whippoorwills. But it is a pair of cardinals who have ended up with the prize piece of real estate—the spot with the bird feeder. I have tapes of birdsongs, and when I play them I try to skip the one of the cardinal, because the current resident goes into a frenzy of territorial song when he hears his rival. His otherwise lovely day is ruined.

And what about the coyote? For a while she was confident that this was her farm, especially the chicken part of it. She was so sure of herself that once she sauntered by in daylight and picked up the tough old rooster to take back to her pups. However, the dogs grew wise to her, and the next few times she returned to exercise her rights they chased her off, explaining that this farm belonged to them and that the chicken flock was their responsibility.

When I start thinking about it that way—that those who inhabit the land and use it have a real claim to it in a nonlegal sort of way—the whole question gets complicated.

A long time ago, before I came to live in the Ozarks, I spent a springtime working on a plot of university research land. I was young and in love, and most tasks seemed happy ones, but the project would have captured my fancy anyway. There were three contrasting habitats being studied: upland forest, bottomland and sandy waste. My job was to dig up a cube of earth from each place every week, sift it, count and rough-classify the inhabitants visible to the naked eye, and then plot the population growth. The resulting curve, a joyous, vibrant freshening of life, matched the weather and my own pulse beat.

That particular love has quieted, and I have not excavated cubes of earth on this place, but I know what is going on down there: Millions of little bodies are fiercely metabolizing and using the land. I dare not even think what numbers I would come up with if I added a pocket lens or microscope to my census-taking tools. But there are other residents I *can* count who do have arguable title here. There are twenty hives of bees back by the woodlot in my home beeyard, each hive containing some 60,000 bees. That makes 1,200,000 bee souls flitting about making claim to all the flowers within two miles.

On the other hand, there are the copperheads, who make walking the fields a boot affair, and all their snakish kin. How am I to count them and judge their claims? There are the turtles who eat the strawberries in the garden, the peepers who own the pond. What about raccoon and skunk and deer rights? What about the bobcat who denned in the cliff by the river and considers my place to be the merest sliver of her own?

It begins to make me dizzy even trying to think of taking a census of everybody who lives here; and all of them seem to have certain claims to the place that are every bit as good as and perhaps better than mine.

Up the road there is a human squabble going on over some land less happily situated. Rather than lying between two environmentally benign government stake-outs, that land and all that surrounds it is in private hands. One owner wants to bulldoze and develop, and so the boundary question is becoming a sticky one. There is talk about having an expensive survey made to establish who owns what. As a spinoff, I suppose that corners will be set and lines run, and then I may know whether this farm is a hundred and five acres or ninety or some other definity.

The indigo buntings probably won't care.

I met Paul, the boy who was to become my husband, when he was sixteen and I was fifteen. We were married some years later, and the legal arrangement that is called marriage worked well enough while we were children and while we had a child. But we grew older, and the son went off to school, and marriage did not serve as a structure for our lives as well as it once had. Still, he was the man in my life for all those years. There was no other. So when the legal arrangement was ended, I had a difficult time sifting through the emotional debris that was left after the framework of an intimate, thirty-year association had broken.

I went through all the usual things: I couldn't sleep or eat, talked feverishly to friends, plunged recklessly into a destructive affair with a man who had more problems than I did but who was convenient, made a series of stupid decisions about my honey business and pretty generally botched up my life for several years running. And for a long, long time, my mind didn't work. I could not listen to the news on the radio with understanding. My attention came unglued when I tried to read anything but the lightest froth. My brain spun in endless, painful loops, and I could neither concentrate nor think with any semblance of order. I had always rather enjoyed having a mind, and I missed mine extravagantly. I was out to lunch for three years.

I mused about structure, framework, schemata, system, classification and order. I discovered a classification Jorge Luis Borges devised, claiming that

> *A certain Chinese encyclopedia divides animals into:*
> *a. Belonging to the Emperor*
> *b. Embalmed*
> *c. Tame*
> *d. Sucking pigs*
> *e. Sirens*
> *f. Fabulous*
> *g. Stray dogs*
> *h. Included in the present classification*
> *i. Frenzied*
> *j. Innumerable*
> *k. Drawn with a very fine camel-hair brush*
> *l. Et cetera*
> *m. Having just broken the water pitcher*
> *n. That from a long way off look like flies.*

Friends and I laughed over the list, and we decided that the fact that we did so tells more about us and our European, Western way of thinking than it does about a supposed Oriental world view. We believe we have a more proper concept of how the natural world should be classified, and when Borges rumples that concept it amuses us. That I could join in the laughter made me realize I must have retained some sense of that order, no matter how disorderly my mind seemed to have become.

My father was a botanist. When I was a child he reserved Saturday afternoons for me, and we spent many of them walking in woods and rough places. He would name the plants we came upon by their Latin binomials and tell me how they grew. The names were too hard for me, but I did understand that plants had names that described their relationships one to another and found this elegant and interesting even when I was six years old.

So after reading the Borges list, I turned to Linnaeus. Whatever faults the man may have had as a scientist, he gave us a beautiful tool for thinking about diversity in the world. The first word in his scheme of Latin binomials tells the genus, grouping diverse plants which nevertheless share a commonality; the second word names the species, plants alike enough to regularly interbreed and produce offspring like themselves. It is a framework for understanding, a way to show how pieces of the world fit together.

I have no Latin, but as I began to botanize, to learn to call the plants around me up here on my hill by their Latin names, I was diverted from my lack of wits by the wit of the system.

Commelina virginica, the common dayflower, is a rangy weed bearing blue flowers with unequal sepals, two of them showy and rounded, the third hardly noticeable. After I identified it as that particular *Commelina,* named from a sample taken in Virginia, I read in one of my handbooks, written before it was considered necessary to be dull to be taken seriously:

Delightful Linnaeus, who dearly loved his little joke, himself confesses to have named the day-flowers after three brothers Commelyn, Dutch botanists, because two of them—commemorated in the showy blue petals of the blossom—published their works; the third, lacking application and ambition, amounted to nothing, like the third inconspicuous whitish third petal.

There is a tree growing in the woodland with shiny, oval leaves that turn brilliant red early in the fall, sometimes even at summer's end. It has small clusters of white flowers in June that bees like, and later blue fruits that are eaten by bluebirds and robins. It is one of the tupelos, and people in this part of the country call it black-gum or sour-gum. When I was growing up in Michigan I knew it as pepperidge. Its botanic name is *Nyssa sylvatica. Nyssa* groups the tupelos, and is derived from the Nyseides—the Greek nymphs of Mount Nysa who cared for the infant Dionysus. *Sylvatica* means "of the woodlands." *Nyssa sylvatica,* a wild, untamed name. The trees, which are often hollow when old, served as beehives for the first American settlers, who cut sections of them, capped them and dumped in the swarms that they found. To this day some people still call beehives "gums," unknowingly acknowledging the common name of the tree. The hollow logs were also used for making pipes that carried salt water to the salt works in Syracuse in colonial days. The ends of the wooden pipes could be fitted together without using iron bands, which would rust.

This gives me a lot to think about when I come across *Nyssa sylvatica* in the woods.

I botanized obsessively during that difficult time. Every day I learned new plants by their Latin names. I wandered about the woods that winter, good for little else, examining the bark of leafless trees. As wildflowers began to bloom in the spring, I carried my guidebooks with me, and filled a fat notebook as I identified the plants, their habitats, habits and dates of blooming. I had to write them down, for my brain, unaccustomed to exercise, was now on overload.

One spring afternoon, I was walking back down my lane after getting the mail. I had two fine new flowers to look up when I got back to the cabin. Warblers were migrating, and I had been watching them with binoculars; I had identified one I had never before seen. The sun was slanting through new leaves, and the air was fragrant with wild cherry (*Prunus serotina: Prunus*—plum, *serotina*—late blooming) blossoms, which my bees were working eagerly. I stopped to watch them, standing in the sunbeam. The world appeared to have been running along quite nicely without my even noticing it. Quietly, gratefully, I discovered that a part of me that had been off somewhere nursing grief and pain had returned. I had come back from lunch.

Once back, I set about doing all the things that one does when one returns from lunch. I cleared the desk and tended to the messages that others had left. I had been gone for a long time, so there was quite a pile to clear away before I could settle down to the work of the afternoon of my life, the work of building a new kind of order, a structure on which a fifty-year-old woman can live her life alone, at peace with herself and the world around her.

One spring evening a couple of years ago, I was sitting in the brown leather chair in the living room reading the newspaper and minding my own business when I became aware that I was no longer alone.

Looking up, I discovered that the three big windows that run from floor to ceiling were covered with frogs.

There were hundreds of them, inch-long frogs with delicate webbed feet whose fingerlike toes ended in round pads that enabled them to cling to the smooth surface of the glass. From their toe structure, size and light-colored bellies, I supposed them to be spring peepers, *Hyla crucifer,* and went outside for a closer look. I had to be careful where I put my feet, for the grass in front of the windows was thick with frogs, waiting in patient ranks to move up to the lighted surface of the glass. Sure enough, each pinkish-brownish frog had a back criss-crossed with the dark markings that give the species its scientific name. I had not known before that they were attracted to light.

I let my newspaper go and spent the evening watching them. They did not move much beyond the top of the windows, but clung to the glass or the moldings, seemingly unable to decide what to do next. The following morning they were gone, and I have never seen them at the windows since. It struck me as curious behavior.

These window climbers were silent; we usually are only aware of spring peepers at winter's end—I first hear their shrill bell-like mating calls in February from the pond up in the field. The males produce the calls by closing their mouths and nasal openings and forcing air from their lungs over the vocal cords into their mouths, and then back over the vocal cords into the lungs again. This sound attracts the females to the pond, and when they enter the water the males embrace them, positioning their vents directly above those of the females. The females then lay their eggs, which the males fertilize with their milt.

It is a clubby thing, this frog mating, and the frogs are so many and their calls so shrill and intense that I like to walk up to the pond in the evenings and listen to the chorus, which, to a human,

is both exhilarating and oddly disturbing at close range. One evening I walked there with a friend, and we sat by the edge of the pond for a long time. Conversation was inappropriate, but even if it had not it would have been impossible. The bell-like chorus completely surrounded us, filled us. It seemed to reverberate with the shrill insistence of hysteria, driving focused thought from our heads, forcing us not only to hear sound but to feel it.

Comparing notes as we walked back to my cabin, we were

startled to discover that we had both wondered, independently, whether that was what it was like to go mad.

A slightly larger cousin of the spring peeper that belongs to the same genus, the gray tree frog, commonly lives in my bee-hives during the summer months. These frogs cling under the protective overhang of the hive cover, and as I pry up the lid, they hop calmly to the white inner cover and sit there placidly eying me.

They are a pleasing soft grayish-green, marked with darker moss-colored patches, and look like a bit of lichen-covered bark when they are on a tree. Having evolved this wonderfully successful protective coloration, the safest behavior for a gray tree frog in a tight spot is to stay still and pretend to be a piece of bark. Sitting on the white inner cover of the beehive, the frog's protective coloration serves him not at all, but of course he doesn't know that, and not having learned any value in conspicuously hopping away, he continues to sit there looking at me with what appears to be smug self-satisfaction and righteous spunk.

Last evening I was reading in bed and felt rather than heard a soft plop on the bed next to me. Peering over the top of my glasses, I saw a plump, proud gray tree frog inspecting me. We studied each other for quite a time, the gray tree frog seemingly at ease, until I picked him up, carried him out the back door and put him on the hickory tree there. But even in my cupped hands he moved very little, and after I put him on the tree he sat quietly, blending in beautifully with the bark. A serene frog.

The sills in my bedroom are rotten, so I supposed that he had found a hole to come through and wondered if he'd had friends. I looked under the bed and discovered three more gray tree frogs, possibly each one a frog prince. Nevertheless, I transferred them to the hickory.

There was something in the back of my mind from childhood Sunday-school classes about a plague of frogs, so I took down my Bible and settled back in bed to search for it. I found the story in Exodus. It was one of those plagues that God sent to convince the Pharaoh to let the Jews leave Egypt.

And the Lord spake unto Moses, Go unto Pharaoh, and say unto him, Thus saith the Lord, Let my people go, that they may serve me.

And if thou refuse to let them go, behold, I will smite all thy borders with frogs:

And the river shall bring forth frogs abundantly, which shall go up and come into thine house, and into thy bedchamber, and upon thy bed . . .

This was exciting stuff; my evening had taken on a positively biblical quality. I was having a plague of frogs, and had obviously had another the evening that the spring peepers had crawled up the living-room windows. Actually, I enjoyed both plagues, but Pharaoh didn't. The writer of Exodus tells us that Pharaoh was so distressed by frogs in his bed that he called Moses and said,

Intreat the Lord, that he may take away the frogs from me, and from my people; and I will let the people go, that they may do sacrifice unto the Lord.

A fussy man, that Pharaoh, and one easily unnerved.

I once knew a pickerel frog, *Rana palustris,* frog of the marshes, who might have changed Pharaoh's mind. The pickerel frog was an appealing creature who lived in my barn one whole summer. He was handsome, grayish with dark, square blotches highlighted with yellow on his legs. I found him in the barn one morning trying to escape the attentions of the cat and the dogs. At some point he had lost the foot from his right front leg, and although the stump was well healed, his hop was awkward and lopsided. I decided that he would be better off taking his chances with wild things, so I carried him out to the pond and left him under the protective bramble thicket that grows there. But the next day he was back in the barn, having hopped the length of a football field to get there. So I let him stay, giving him a dish of water and a few dead flies.

All summer long I kept his water fresh, killed flies for him and kept an eye out for his safety. Pickerel frogs sometimes live in caves, and I wondered if the dim light of the barn and the cool concrete floor made him think he had discovered a cave where the service was particularly good. That part of the barn serves as a passageway to my honey house, and I grew accustomed to seeing him as I went in and out of it. I came to regard him as a tutelary sprite, the guardian of the honey house, the Penate Melissus.

Then one day the health inspector came for his annual tour. Like Pharaoh, the health inspector is a fussy man. Once he gave me a hard time because there were a few stray honeybees in the honey house. Bees, he explained patiently, were insects, and regulations forbade insects in a food-processing plant. I pointed out, perhaps not so patiently, that these insects had made the food, and that until I took it from them, they were in continuous, complete and intimate contact with it. He gave up, but I know he didn't like it. So I wasn't sure how he'd react to the pickerel frog squatting outside the honey-house door with his bowl of water and mason-jar lid full of dead flies. But the health inspector is a brisk man, and he walked briskly by the frog and never saw him. I was thankful.

Years ago, in an introductory biology class, I cut up a frog, carefully laying aside the muscles, tracing the nerves and identifying the organs. I remember that as I discarded the carcass I was quite pleased with myself, for now I knew all about frogs and could go on to learn the remaining one or two things about which I still had some small ignorance. I was just about as smug as a gray tree frog on a white beehive.

In the years after that, and before I moved to the Ozarks, I also lived a brisk life, and although I never had much reason to doubt that I still knew all about frogs, I don't think I ever thought about them, for, like the health inspector, I never saw any.

Today my life has frogs aplenty and this delights me, but I am not so pleased with myself. My life hasn't turned out as I expected it would, for one thing. For another, I no longer know all about anything. I don't even know the first thing about frogs, for instance. There's nothing like having frogs fill up my windows or share my bed or require my protection to convince me of that.

I don't cut up frogs anymore, and I read more poetry than I did when I was twenty. I just read a couplet about the natural world by an anonymous Japanese poet. I copied it out and put it up on the wall above my desk today:

Unknown to me what resideth here
Tears flow from a sense of unworthiness and gratitude.

My three hundred hives of bees are scattered across the hills of southern Missouri in outyards on farmers' pastures or at the edge of their woodlots. I give each family who has one of those outyards a gallon of honey a year as rent, but mostly farmers just like having the hives around, for the bees pollinate their fruit and vegetables and the clover in their pastures.

Bees fly two miles or more as they forage for nectar. I once calculated that the eighteen million bees in my hives cover a thousand square miles of the Ozarks in their flights. In the spring I spend most of my time driving to the outyards, taking care of the bees and getting them ready for the major nectar flows from which they will make the honey crop.

I use a big truck to haul the honey to market, but for the work around here I drive a courageous 1954 red Chevy half-ton pickup named "Press on Regardless" because it runs zestfully without many parts that are normally considered to be automotive necessities. That pickup and I have been over some rough spots, through mud and hard times, and I like to take good care of it.

Today was cold and rainy, not good weather for bee work, so I spent the day working on the pickup. I went out to the center part of the barn where I keep it, made a fire in the wood stove and turned the radio on to the public station I can get from the university eighty miles to the north. The radio promised that it would give me Albinoni's Concerto in C for Two Oboes and Strings. I started checking the pickup's vital fluids. The transmission oil was okay, but I had to add half a can of brake fluid to the master cylinder. I let out the crankcase oil, as it was time to change it. While the oil was draining, I jacked up the front of the pickup, rolled under it on my creeper and with bits of caked mud and grease falling into my eyes, greased the front end. My pickup has twenty grease points, and they have to be tended often in a truck that old.

While lying there on the creeper, Albinoni drifting through my brain, I noticed that I had been careless with the right rear shock absorber. Not only was it missing, but the bracket that held it in place had ripped loose. I added five quarts of new oil to the crankcase, reversed the truck, checked the oil level in the rear end, and then jacked it up to grease the rear springs. Afterwards I replaced the starter button above the starter assembly. The foot starter has been balky recently, and Ermon, my neighbor, a mechanic who lives across the hollow, had told me that the button needed replacing and showed me how to do it. The old one came out easily enough, but screwing in the new one with my fingertips backed up against the engine block while I sprawled over the fender was hard and took me past Albinoni, through Mozart, Beethoven and right on up to the Romantics before I was done. I wondered how the mechanic with his bigger fingers could manage it so deftly.

After I was finished, I peeled off my greasy coveralls, pleased and mildly surprised as I always am to find myself clean enough underneath to receive company in a parlor.

Outdoors the weather had still not improved for beework, so I drove into town to do errands. I stopped at the auto-parts store to buy a new set of shock absorbers, and then went over to my favorite salvage yard to buy a spare universal joint. The kind of driving that I do is hard on universal joints, and the last time I sheared one in two my neighbor mechanic had replaced it with the last one he had and told me to find another that I could carry as a spare in my glove box.

The salvage-yard man is a friend of mine, and I knew that buying the universal joint would take some time, just the thing that both he and I would enjoy on a rainy day. When I went into his shop he was cutting a junked Pontiac into usable parts with a welding torch. I leaned up against his tool bench to wait for him to finish a cut. When he stopped, he pushed back his welding helmet and nodded to me. He poured a cup of coffee for me and another for himself. We talked about the weather, discussing the pros and cons of rain past, present and future. Then he wondered what I was a-needing. I mentioned the universal joint for my Chevy. No, no, he seldom got those old rigs in nowadays, and they were always stripped clean of good parts when he did.

I have lived in the Ozarks twelve years now, so I did not say thank you and leave. I knew that so far he had just said howdy. I asked him how he was getting along with the 1934 pickup he had bought and was restoring.

"Not getting along worth a damn," he said. "Seems like folks come in looking for parts and interrupting all the time. Got me as fussed as a fart in a mitten." He looked hard at me. I took his comment for what it was worth and settled back more comfortably against the toolbench.

"Want to get to work on it, too," he said. "It's just like the first pickup that I ever did own. Say, did I tell you about that rig?"

No, he hadn't.

He leaned back against a crumpled fender. "It was a thirty-four Chevy that I'd bought from old Peg-Leg Potter," he said. "It was a pretty thing, and I was proud to drive that pickup home. But the next day I noticed that it was making a knocking noise that just wasn't right at all, so I crawled under that pickup and took the oil pan off. Well, old Peg-Leg Potter, whatever else he was, he wasn't much of a mechanic, because he'd worn out a bearing and just wrapped a piece of bacon rind round the rod to make do. Why, you cain't run no pickup on bacon rind! Well, I caught my daddy asleep one day and I cut me a piece of shoe leather right out of the tongue of his shoe, and I took out that bacon rind and put the shoe leather in its place, and then I drove that thirty-four Chevy on down the road and traded it for a 1948 Ford. The old boy that I traded to, he come round to see me the next week complaining about the shoe leather. I says to him, I says, 'Don't talk to me about it. Why, I *improved* that pickup. You go talk to old Peg-Leg Potter. He thinks you can run a pickup on bacon rind.' "

If I had laughed or even smiled I would have ruined the transaction, so I tried to arrange my face in a way that acknowledged that I'd heard a good story but that I still knew a joke from a jaybird and that I still lacked a universal joint. I said that the rain appeared to be letting up some and I guessed I'd get going.

"If you wasn't in such an all-fired hurry, I expect I might find you a U-joint somewheres," he said, and rummaged around on a shelf until he found one.

The only time I have ever wished I were a man was at that moment, for what I should have said next was, "How much would a man have to give for that old U-joint now?" But I'm not and I couldn't and I didn't. And the word "woman" won't work in that question. So instead I asked him how much it was, he told me, I paid him and he returned to his cutting torch.

Spring 23

Anyone who has kept bees is a pushover for a swarm of them. We always drop whatever we are doing and go off to pick one up when asked to do so. It doesn't make sense, because from a standpoint of serious beekeeping and honey production a swarm isn't much good. Swarms are headed up by old queens with not much vitality or egg-laying potential left, and so a beekeeper should replace her with a new queen from a queen breeder. He will probably have to feed and coddle the swarm through its first year; it will seldom produce any extra honey the first season. And yet we always hive them.

There is something really odd about swarms, and I notice that beekeepers don't talk about it much, probably because it is the sort of thing we don't feel comfortable about trying to put into words, something the other side of rationality.

The second year I kept bees, I picked up my first swarm. I was in the middle of the spring beework, putting in ten to twelve hours a day, and very attuned to what the bees were doing out there in their hives. That day had begun with a heavy rainstorm, and so rather than working out in the beeyards, I was in the honey house making new equipment. By afternoon the rain had stopped, but the air was warm and heavy, charged and expectant. I began to feel odd, tense and anticipatory, and when the back of my neck began to prickle I decided to take a walk out to the new hives I had started. Near them, hanging pendulously from the branch of an apple tree, was a swarm of bees. Individual bees were still flying in from all directions, adding their numbers to those clinging around their queen.

In the springtime some colonies of bees, for reasons not well understood, obey an impulse to split in two and thus multiply by swarming. The worker bees thoughtfully raise a new queen bee for the parent colony, and then a portion of the bees gather with the old queen, gorge themselves with honey and fly out of the hive, never to return, leaving all memory of their old home behind. They cluster somewhere temporarily, such as on the branch of my apple tree. If a beekeeper doesn't hive them, scout bees fly from the cluster and investigate nearby holes and spaces, and report back to the cluster on the suitability of new quarters.

We know about two forms of honeybee communication. One is chemical: information about food sources and the wellbeing of the queen and colony is exchanged as bees continually feed one another with droplets of nectar which they have begun to process and chemically tag. The other form of communication is tactile: bees tell other bees about good things such as food or the location of a new home by patterned motions. These elaborate movements, which amount to a highly stylized map of landmarks, direction and the sun's position, are called the bee dance.

Different scout bees may find different locations for the swarm and return to dance about their finds. Eventually, sometimes after several days, an agreement is reached, rather like the arrival of the Sense of the Meeting among Quakers, and all the bees in the cluster fly off to their new home.

I watched the bees on my apple tree for a while with delight and pleasure, and then returned to the barn to gather up enough equipment to hive them. As I did so, I glanced up at the sky. It was still dark from the receding thunderstorm, but a perfect and dazzling rainbow arched shimmering against the deep blue sky, its curve making a stunning and pleasing contrast with the sharp inverted V of the barn roof. I returned to the apple tree and shook the bees into the new beehive, noticing that I was singing snatches of one of Handel's coronation anthems. It seemed as appropriate music to hive a swarm by as any I knew.

Since then, I have learned to pay attention in the springtime when the air feels electric and full of excitement. It was just so one day last week. I had been working quietly along the row of twelve hives in an outyard when the hair on the back of my neck began to stand on end. I looked up to see the air thick with bees flying in toward me from the north. The swarm was not from any of my hives, but for some reason bees often cluster near existing hives while they scout a new location. I closed up the hive I was working on and stood back to watch. I was near a slender post oak sapling, and the bees began to light on one of its lower limbs right next to my elbow. They came flying in, swirling as they descended, spiraling around me and the post oak until I was enveloped by the swarm, the air moving gently from the beat of their wings. I am not sure how long I stood there. I lost all sense of time and felt only elation, a kind of human emotional counterpart of the springlike, optimistic, burgeoning, state that the bees were in. I stood quietly; I was nothing more to the bees than an object to be encircled on their way to the spot where they had decided, in a way I could not know, to cluster. In another sense I was not remote from them at all, but was receiving all sorts of meaningful messages in the strongest way imaginable outside of human mental process and language. My skin was tingling as the bees brushed past and I felt almost a part of the swarm.

Eventually the bees settled down in the cluster. Regaining a more suitable sense of my human condition and responsibilities, I went over to my pickup and got the empty hive that I always carry with me during swarming season. I propped it up so that its entrance was just under the swarm. A frame of comb from another hive was inside and the bees in the cluster could smell it, so they began to walk up into the entrance. I watched, looking for the queen, for without her the swarm would die. It took perhaps twenty minutes for all of them to file in, and the queen, a long, elegant bee, was one of the last to enter.

I screened up the entrance and put the hive in the back of the pickup. After I was finished with my work with the other hives in the beeyard, I drove back home with my new swarm.

I should have ordered a new queen bee, killed the old one and replaced her, but in doing that I would have destroyed the identity of the swarm. Every colony of bees takes its essence, character and personality from the queen who is mother to all its members. As a commercial beekeeper, it was certainly my business to kill the old queen and replace her with a vigorous new one so that the colony would become a good honey producer.

But I did not.

The local VFW has a campground on the river directly below my farm. During the warm weather, its members and their families come there to swim, cook out and sometimes camp overnight.

The VFW is an important social organization in town and sponsors civic events as well as fish fries, barbecues and pig roasts at the slightest hint of a patriotic occasion. The members are mostly World War II veterans. Here as elsewhere, many of the veterans of Vietnam are a bitter lot, not interested in socializing in an organization reminding them of war. One of them, who was repairing a tire for me at a gas station in town, asked if I wasn't the Bee Lady who lived up by the VFW campground? I was. Was he a member of the VFW?

"Nah, I can't stand listening to those old guys sitting around telling about how a war should be fought," he said savagely.

None of the VFW members or their wives are close friends of mine, but I know many in the group and they often invite me to join them at their cookouts. The invitations are a friendly gesture, not meant to be accepted, but a politeness that establishes that we are good neighbors. But, like my mother, the VFW seems to believe that I don't eat properly, because every time they cook out down there, one of the veterans drives back up the hill with a box full of barbecued meat, fried potatoes, baked beans, and salad, enough food for days.

"Brought you some groceries," the veteran will say with a grin, and drive back down the hill to his friends.

At the end of the summer, the VFW holds a stag party on a Saturday. There is very little real hell-raising from what I ever hear, but much talk of it beforehand. One member told me that the wife of a new recruit was upset because they had told her that on the night of the stag party a canoe full of naked women always floated down the river at precisely 10 P.M. The man's wife was just a young thing and she believed them and got pretty riled up, the veteran told me, laughing hard, and then added, "Hell, by ten o'clock we're all so drunk we wouldn't know if the Queen of Sheba floated down the river."

One afternoon last summer, Virgil, a VFW member, and his wife, Mary Lou, whom I know well, stopped by my place and said they were roasting a goat down there for a few friends and really would like me to join them. I hadn't seen them for a while and had been working hard, so the prospect of an evening of roasted goat and good-old-boy banter was a welcome one; I thanked them and said I'd come down later. Before I left I searched for something to contribute to the feast. The refrigerator held some scraps of raw vegetables, just enough for the salad I had intended to have for supper, an opened jar of mustard that was evolving into a higher life form, and half a jug of red wine left over from dinner with friends the week before. The cupboard was no better: some stale crackers and half a bottle of teriyaki sauce. Half a jug of wine seemed more festive than half a bottle of teriyaki sauce, so I slung it over my shoulder and walked down the hill to the campground. When I got there, Virgil peered at the level of wine in the jug.

"I knew it was a long walk down the hill, but I didn't know it was *that* long," he said, his face innocent of a smile.

The evening was a pleasant one. The goat, roasted in herbs, was delicious, the company good, and Virgil kept everyone limp with laughter with his straight-faced stories. I was glad I had accepted their invitation.

Last Friday night a tired and troubled veteran shot and killed himself at the campground. I did not hear the shot and did not know the man, but I did know the two veterans who were pounding on my door minutes later.

I was in bed reading when I heard them. I got up, put on a robe, and answered the door. The two men were incoherent and weeping. I knew something was terribly wrong but I couldn't find out what. I brought them into the living room and had them sit down, and gradually made out what had happened. They had been trying for two hours to talk the man out of suicide, but he had gone ahead with it and they were racking themselves for their failure.

They wanted to use my telephone to call the sheriff and the man's family, but the horror still in front of their eyes made it impossible for them to read the telephone book. I looked up the numbers for them, helped them to place their calls and tried to find something to say, but there were no words that could undo what had been done. Then they asked me to call Virgil, a good man in a crisis. They wanted him with them. I said I would, and asked them to stay with me for the half hour that it would take the sheriff to get here. I would make some coffee. I did not think it was good for them to go back and sit alone with that body whose head had been ripped apart with a shotgun blast, but they would not stay, and drove their pickup down to the river, the wheels spinning gravel in the darkness. I made the call to Virgil, and he said he would go to them immediately.

Then my telephone began to ring. Here in the country, listening to the police scanner vies with television as an evening's entertainment. People throughout the neighborhood had heard the sheriff saying he was on his way to investigate a reported death at the VFW campground, and the curious were calling to ask what had happened.

I cannot see the road from my cabin, but soon I heard traffic on it and knew that the sheriff, Virgil and perhaps some of the other veterans were arriving; I felt a little easier about the two men who had been here. Sleep was impossible that night, and I thought with sorrow of the unknown man whose life had become such pain to him that he had to leave it.

The next afternoon, one of the veterans who had knocked on my door the night before came to see me. His eyes were red, his face was creased with fatigue and he was staggeringly drunk. He told me that all the boys had decided the best thing to do was to have a family party at the campground for the rest of the weekend and to camp out there that night, or else no one would ever be able to go there again. It was important that I come down for dinner. It was important. Very. Very. Important. Very important. Very.

I said I would come, and assured him that I understood.

The campground had been taken over by death; it had to be returned to life and parties before it grew a ghost. The two men needed to talk to me at a party there. The night before they had not been just a pair of good old boys, but two men who had seen horror and had brought it into my living room. I had not been the Bee Lady on the hill, but a woman who had put her arms around them when they cried. Now it was time for us to go back to what we had all been before.

When the sun set, I walked down the hill to the campground. We all ate something or other. I sat with the circle of wives and daughters, and we talked of something or other. The men sat by themselves, drinking. After a while I talked with the two men who had come to my cabin. We agreed that what had happened was bad, but that time would help a person to forget. They thanked me for letting them use the telephone, and then I walked back up the hill.

SUMMER

I am an early riser, and now that the weather is warm I like to take a cup of coffee out under the oak trees in back of the cabin and get a feel for the kind of day it is going to be. Today the night creatures were still about when I went out there—katydids, whippoorwills, night-flying moths, owls and mosquitoes. By the time I had had a sip or two of coffee and my eyes had adjusted well enough to pick out the shapes of the trees, the mosquitoes had discovered me and gathered in an annoying buzz around my head. But before they had a chance to bite, a small furry shape appeared from nowhere. I heard the soft rush of wings beside my ear and the mosquitoes were gone. A few moments of silence. More mosquitoes, and once again a bat swooped in.

The arrangement was a pleasant one for both the bat and me. I don't like mosquitoes but the bat does. I served as bait to gather the mosquitoes in one rich spot, and the bat ate them before they bit me. For the mosquitoes the plan was not such a good one; they were kept from dinner and were turned into one instead. All this gives me a fine, friendly feeling toward bats. In their way, I suppose, they also approve of me.

The bats are quick, and in the dim light before dawn it is difficult to identify them, but I believe them to be *Myotis lucifugus,* more comfortably known as little brown bats. These, at least, are the ones I often see taking their daytime sleep hanging upside down from the rafters in the barn loft. They are common around here, and also sleep in caves or hollow trees. Like other bats, they belong to the order Chiroptera, which means wing-handed, a good word for an animal with wings made of a membrane of skin covering the hand and fingerbones. But I also like the old English name of flittermouse, an apt description of our only flying mammal.

Bats are mammals like we are. They suckle their young, and have such wizened ancient-looking faces that they seem strangely

akin and familiar. Yet they find their way and locate their food by using sound that we cannot hear. They hunt by night, and in cold weather some migrate and others hibernate. They are odd and alien to us, too, so much so that we have made up fancies about them—that they are evil and ill-omened, or at the very least will fly into our hair. Anyone who has read *Dracula* will remember that young ladies should not moon around graveyards at night, or they will be in big trouble with bats.

The truth is that, from a human point of view, bats are beneficial. On the North American continent, the little brown bat and other temperate-zone bats have a diet made up almost exclusively of night-flying insects. Farther south, in tropical regions, there are bats that eat fruit, and even vampire bats, which have incisors that allow them to feed on the blood of large animals, but our northern bats have nothing to do with such fare, eating, instead, the kinds of insects that are often a bother to us.

Bats find their food by producing high-pitched sounds that echo against insects or other solid objects. The echos return to the bats' ears and give them a precise and vivid aural map of whatever is out there.

These cries are beyond our human range of hearing, but when translated to frequencies low enough for us they sound like a series of short clicks or chirps. The short ultrasound wave-lengths allow bats to locate exactly even very small moving objects, such as mosquito-sized insects. They pluck the mosquitoes out of the air close to my head, and would never be so clumsy as to blunder into my hair.

What is more, the bats' discrimination is so fine that they can separate the echos of their own clicks from those of another bat. This is important, because they often fly and hunt in groups, and such accuracy allows them to do so without muddle or confusion. It may not be our way to get dinner, but it does strike me as wonderfully clever, efficient and simple. However, as with most things in life, it is not all that simple, and dinner is not always a sure thing.

Night-flying moths are one of the chief items in bat diet, and over a long time span of eating and being eaten, bat and moth have worked out a complex relationship.

My cousin Asher, whose academic specialty is moth ear mites, has during the course of a lifetime of work discovered a good deal about moths' ears, a structure that some of us never knew existed. He tells me that some moths who are night fliers and therefore potential bat dinners can hear the high-pitched noises made by bats and stand a good chance of escaping. What is even more remarkable is that certain moths have the ability to make sounds that the bats, in turn, can hear. In a sense, they can talk back to bats.

"And what," I asked Asher, "do they say?"

"They say, 'I am not good to eat,'" he replied.

This makes life harder for the bats and easier for the moths, but the moths' advantage is only preserved by a highly specialized relationship with yet another creature.

The moth ear mites Asher studies, the North American species, harm their hosts' ears, and if they were not careful, they would deafen them and make the moths and themselves easy prey. But they *are* careful.

The mites are tiny arachnids, scarcely visible to the naked eye. When they are ready to lay their eggs, they climb on the moth and make their way to the moth's ear, a safe and protected spot for their eggs. In the process of laying the eggs, they damage the delicate structure of the moth's ear. Since many mites may be present on a single moth, the moth would be deaf if they were to lay their eggs in both ears. So, in a stunning example of evolutionary respect, a case where courtesy and self-interest are one and the same, the first mite aboard makes a trail, in a manner not yet clearly understood, and all the mites who come later follow her trail, laying all their eggs in the same ear and leaving the opposite ear undamaged. This allows the moth to retain partial hearing, and may improve his chances of escaping bats during the time the mites' eggs hatch.

So there we are out under the oak trees in the dim light—the mites, the moths, the bats, the mosquitoes, and me. We are a text of suitability one for another, and that text is as good as any I know by which to drink my coffee and watch the dawn.

I was bitten by a brown recluse spider a couple of weeks ago and lived to tell about it, so I shall.

Brown recluses are the most poisonous spiders in the United States. Unlike black widows, the bite of both sexes is venomous; also unlike black widows, they are common in houses throughout the Southeast and Southwest.

The one that bit me was hiding between the folds of a towel that I picked up and flung over my shoulder as I got ready to go for a swim with a friend. We were walking along the pathway down to the river when I felt a sharp, burning bite on my upper arm. I dropped the towel and saw a brown recluse scurry away. They are easy to recognize: grayish-brown leggy spiders half an inch or so in size, with the distinctive brown violin-shaped marking on the top of their broad cephalothorax that gives them their other common name—violin or fiddle spider. My friend was horrified when I told him it was a brown recluse that had bitten me. He asked if I wanted to go back to the cabin. "Why?" I said. "If I'm going to die, I might better die down at the river than back in the cabin." So we continued on our way and spent the afternoon swimming and lying on a gravel bank.

I died hardly at all, and the bite never amounted to much, nothing like one from a tick or a mosquito. Within a few days the pimple-like mark had disappeared without a trace. Many people—most, as a matter of fact—have a reaction no worse than mine to the bite of a brown recluse, but some lack immunity to the venom, and although death from the bite is extremely rare, they can have a severe reaction. The spider's enzymes produce neurotoxins that break down cell membranes, destroying blood vessels and causing clotting. The surrounding tissue dies, creating an ever-widening wound that is so stubborn in healing that in exceptional cases skin grafts are required to repair the damage. In addition, a susceptible person may experience chills, nausea and fever. The reaction is highly variable, but fortunately can be medically treated.

Although they spin untidy webs under rocks and logs, brown recluses are for the most part common indoor spiders. Their scientific name, *Loxosceles reclusa* (*Loxosceles* means slant-legged), dubs them reclusive, but although they may be solitary, they can be seen frequently throughout the house. To be sure, they prefer to hide in the folds of clothes in a drawer or in towels on the

shelf, as did the one who bit me, but they come out to explore and forage for small crawling insects. They cannot climb smooth surfaces, and I often find them trapped in the bathtub or sink, skittering about trying to escape. A friend and her daughter stopped in for tea not long ago. I made the tea in a pot and handed out cups. Accustomed to country living, the daughter wisely peered into her cup before I poured the tea. "Hmm! A brown recluse," she said calmly, and we dumped out the spider and rinsed the cup before we had our tea party.

Most people are not as unflappable as my young friend. Descriptions of skin rotting off from terrible brown recluse bites circulate from time to time, and are greeted with horrified shudders. The brown recluse is one of our modern monsters. The day after I was bitten, I read in one of my spider books something that at first surprised me: Although the bites of other brown spiders in South America, other species of *Loxosceles,* were known to make people sick, it wasn't until the 1950s, as a result of bites in Texas, Kansas, Missouri and Oklahoma, that the brown recluse was recognized to be similarly toxic. Upon reflection, I realized that it was testimony to how minor the bite is for most people. Humans and brown recluse spiders have been sharing living quarters for a long time, and people must have been bitten now and again, but it wasn't until the occasional toxicity of the bite was established that the creature was turned into a Scary Beast.

A newspaper story appeared in several cities recently about the development of an antidote for brown recluse bites. After reading the story, a city friend telephoned me and asked if I'd ever seen a brown recluse. I told her that they were common, and now she refuses to visit me, although she has been here a number of times and has returned home in perfect health after each visit. She is a good friend and I shall miss her company.

Last week I was in St. Louis and went to a party with friends. When some people there learned that I lived in the country, they asked me about brown recluse spiders. Having recently been bitten and read up on the topic, I jumped right in, telling them rather more than they wanted to know about the infrequency and usual mildness of the bites and the shy nature of the little spider. What they wanted to hear more about was the part where the skin rots off. After scaring themselves deliciously for a while, several of them decided to cancel plans for a weekend in the Ozarks, and I realized that one of the major points in the favor of brown recluse spiders is that they help keep down the tourists.   *Summer 43*

This week I have started cutting my firewood. It should be cut months ahead of time to let it dry and cure, so that it will burn hot in the winter. It is June now, and almost too late to be cutting firewood, but during the spring I was working with the bees from sunup until sundown and didn't have time. By midday it is stifling back in the woods, so I go out at sunrise and cut wood for a few hours, load it into the pickup and bring it back to stack below the barn.

I like being out there early. The spiders have spun webs to catch night-flying insects, and as the rising sun slants through the trees, the dewdrops that line the webs are turned into exquisite, delicate jewels. The woodlot smells of shade, leaf mold and damp soil. Wild turkey have left fresh bare spots where they scratched away the leaves looking for beetles and grubs. My dogs like being there too, and today snuffled excitedly in a hollow at the base of a tree. The beagle shrieked into it, his baying muffled. The squirrel who may have denned in the tree last night temporarily escaped their notice and sat on a low limb eying the two dogs suspiciously, tail twitching. A sunbeam lit up a tall thistle topped with a luxuriant purple blossom from which one butterfly and one honeybee sipped nectar. Red-eyed vireos sang high in the treetops where I could not see them.

For me their song ended when I started the chain saw. It makes a terrible racket, but I am fond of it. It is one of the first tools I learned to master on my own, and it is also important to me. My woodstove, a simple black cast-iron-and-sheet-metal affair, is the only source of heat for my cabin in the winter, and if I do not have firewood to burn in it, the dogs, cat, the houseplants, the water in my pipes and I will all freeze. It is wonderfully simple and direct: cut wood or die.

When Paul was here he cut the firewood and I, like all Ozark wives, carried the cut wood to the pickup. When he left, he left his chain saw, but it was a heavy, vibrating, ill-tempered thing. I weigh a hundred and five pounds, and although I could lift it, once I had it running it shook my hands so much that it became impossibly dangerous to use. One year I hired a man to cut my wood, but I was not pleased with the job he did, and so the next year, although I could not afford it, I bought the finest, lightest, best-made chain saw money could buy. It is a brand that many woodcutters use, and has an antivibration device built into even its smaller models.

The best chain saws are formidable and dangerous tools. My brother nearly cut off his arm with one. A neighbor who earns his living in timber just managed to kill the engine on his when he was cutting overhead and a branch snapped the saw back toward him. The chain did not stop running until it had cut through the beak of his cap. He was very solemn when I told him that I had bought my own chain saw, and he gave me a good piece of advice. "The time to worry about a chain saw," he said, "is when you stop being afraid of it."

I am cautious. I spend a lot of time sizing up a tree before I fell it. Once it is down, I clear away the surrounding brush before

I start cutting it into lengths. That way I will not trip and lose my balance with the saw running. A dull chain and a poorly running saw are dangerous, so I've learned to keep mine in good shape and I sharpen the chain each time I use it.

This morning I finished sawing up a tree from the place where I had been cutting for the past week. In the process I lost, in the fallen leaves somewhere, my scrench—part screwdriver, part wrench—that I use to make adjustments on the saw. I shouldn't have been carrying it in my pocket, but the chain on the saw's bar had been loose; I had tightened it and had not walked back to the pickup to put it away. Scolding myself for being so careless, I began looking for another tree to cut, but stopped to watch a fawn that I had frightened from his night's sleeping place. He was young and his coat was still spotted, but he ran so quickly and silently that the two dogs, still sniffing after the squirrel, never saw him.

I like to cut the dead trees from my woodlot, leaving the ones still alive to flourish, and I noticed a big one that had recently died. This one was bigger than I feel comfortable about felling. I've been cutting my own firewood for six years now, but I am still awed by the size and weight of a tree as it crashes to the ground, and I have to nerve myself to cut the really big ones.

I wanted this tree to fall on a stretch of open ground that was free of other trees and brush, so I cut a wedge-shaped notch on that side of it. The theory is that the tree, thus weakened, will fall slowly in the direction of the notch when the serious cut, slightly above the notch on the other side, is made. The trouble is that trees, particularly dead ones that may have rot on the inside, do not know the theory and may fall in an unexpected direction. That is the way accidents happen. I was aware of this, and scared, besides, to be cutting down such a big tree; as a result, perhaps I cut too timid a wedge. I started sawing through on the other side, keeping an eye on the tree top to detect the characteristic tremble that signals a fall. I did not have time to jam the plastic wedge in my back pocket into the cut to hold it open because the tree began to fall in my direction, exactly opposite where I had intended. I killed the engine on the saw and jumped out of the way.

There was no danger, however. Directly in back of where I had been standing were a number of other trees, which was why I had wanted to have the sawed one fall in the opposite direction; as my big tree started to topple, its upper branches snagged in another one, and it fell no further. I had sawed completely through the tree, but now the butt end had trapped my saw against the stump. I had cut what is descriptively called a "widow maker." If I had been cutting with someone else, we could have used a second saw to cut out mine and perhaps brought down the tree, but this is dangerous and I don't like to do it. I could not even free my saw by taking it apart, for I had lost my scrench, so I drove back to the barn and gathered up the tools I needed; a socket wrench, chains and a portable winch known as come-along. A come-along is a cheery, sensible tool for a woman. It has a big hook at one end and another hook connected to a steel cable at the other. The cable is wound around a ratchet gear operated by a long handle to give leverage. It divides a heavy job up into small manageable bits that require no more than female strength, and I have used it many times to pull my pickup free from mud and snow.

The day was warming up and I was sweating by the time I got back to the woods, but I was determined to repair the botch I had made of the morning's woodcutting. Using the socket wrench, I removed the bar and chain from the saw's body and set it aside. The weight of the saw gone, I worked free the bar and chain pinched under the butt of the tree. Then I sat down

on the ground, drank ice water from my thermos and figured out
how I was going to pull down the tree.

Looking at the widow maker, I decided that if I could wind
one of the chains around the butt of it, and another chain around
a nearby standing tree, then connect the two with the come-
along, I might be able to winch the tree to the ground. I attached
the chains and come-along appropriately and began. Slowly, with
each pump of the handle against the ratchet gear, the tree sank
to the ground.

The sun was high in the sky, the heat oppressive and my shirt
and jeans were soaked with sweat, so I decided to leave the job
of cutting up the tree until tomorrow. I gathered my tools
together, and in the process found the scrench, almost hidden in
the leaf mold. Then I threw all the tools into the back of the
pickup, and sat on the tailgate to finish off the rest of the ice water
and listen to the red-eyed vireo singing.

It is satisfying, of course, to build up a supply of winter
warmth, free except for the labor. But there is also something
heady about becoming a part of the forest process. It sounds
straightforward enough to say that when I cut firewood I cull
and thin my woods, but that puts me in the business of deciding
which trees should be encouraged and which should be taken.

I like my great tall black walnut, so I have cut the trees around
it to give it the space and light it needs to grow generously.
Dogwoods don't care. They frost the woods with white blossoms
in the spring, and grow extravagantly in close company. If I clear
a patch, within a year or two pine seedlings move in, grow up
exuberantly, compete and thin themselves to tolerable spacing. If
I don't cut a diseased tree, its neighbors may sicken and die. If
I cut away one half of a forked white oak, the remaining trunk
will grow straight and sturdy. Sap gone, a standing dead tree like
the one I cut today will make good firewood, and so invites

cutting. But if I leave it, it will make a home for woodpeckers, and later for flying squirrels and screech owls. Where I leave a brush pile of top branches, rabbits make a home. If I leave a fallen tree, others will benefit: ants, spiders, beetles and wood roaches will use it for shelter and food, and lovely delicate fungi will grow out of it before it mixes with leaf mold to become a part of a new layer of soil.

One person with a chain saw makes a difference in the woods, and by making a difference becomes part of the woodland cycle, a part of the abstraction that is the forest community.

I've been out in back today checking beehives. When I leaned over one of them to direct a puff of smoke from my bee smoker into the entrance to quiet the bees, a copperhead came wriggling out from under the hive. He had been frightened from his protected spot by the smoke and the commotion I was making, and when he found himself in the open, he panicked and slithered for the nearest hole he could find which was the entrance to the next beehive. I don't know what went on inside, but he came out immediately, wearing a surprised look on his face. I hadn't known that a snake could look surprised, but this one did. Then, after pausing to study the matter more carefully, he glided off to the safety of the woods.

He was a young snake, not even two feet long. Like the other poisonous snakes found in the Ozarks, the cottonmouths, copperheads belong to the genus *Agkistrodon,* which means fish-hook toothed. The copperheads in my part of the Ozarks are the southern variety, *Agkistrodon contortrix contortrix,* which makes them sound very twisty indeed. They are a pinkish coppery color with darker hourglass-patterned markings. They have wide jaws, which give their heads a triangular shape. Like the cottonmouths, they are pit vipers, which means that between eye and nostril they have a sensory organ that helps them aim in striking at warm-blooded prey. They eat other small snakes, mice, lizards and frogs.

The surprising thing about copperheads are their mild manners, timidity and fearfulness. They have, after all, a potent defensive weapon in their venom, and yet their dearest wish when they are discovered is to escape. This rocky upland peninsula of land between the river and creek is a lovely habitat for copperheads. I often find them under the beehives, and they are common in the open field. Twice I have had them in the cabin. Every time I come upon copperheads they simply try to get away from me and never offer to strike.

Once I had an old and heavy Irish setter who was badly bitten when he clumsily stepped on a copperhead. To the snake, being walked on by eighty-five pounds of dog represented a direct attack, and so he struck. The dog's leg swelled and he was in obvious pain. Within a few hours his heartbeat was rapid, his breath shallow, and I took him to the vet. Afterwards he watched where he put his feet. I do, too, and wear leather boots when I walk through the field or in the woods, and in the warm months I give decent warning when I turn over a stack of old boards. I have enormous respect for a small animal with venom so potent that it can make a large dog very sick. I weigh more than the dog, and so I might not have such a severe reaction; there is no record of a human death caused by a copperhead bite in Missouri, but I don't want to risk the pain.

I respect copperheads, but I also have another set of feelings toward them, a combination of amazement and sympathy that an animal should be so frightened by me, so eager to escape, so little inclined to use the powerful means that he has to defend himself.

Copperheads contrast oddly with the eastern hognose snakes, sometimes called puff adders, that I also see here sometimes. These are harmless, but put on a tremendous show of ferocity. I came upon a hognose one day in the field, and he raised up the first third of his body and spread his neck wide, hissing horribly, trying to convince me that he was a cobra. I was fooled hardly at all and stood quietly watching him; after some more half-hearted hissing and spreading he gave up the attempt to frighten me, remembered urgent business he had elsewhere and slithered away into the tall grasses.

Apart from copperheads, there are few dangerous snakes here. There are supposed to be rattlers, but in the twelve years I have been walking the woods and river banks I have never met one. Most of the snakes around are harmless or, like the black rat snakes, which eat rodents, beneficial, and I have no sympathy with the local habit of killing every snake in sight. It is an Ozark custom to pack a pistol along with the beer on a float trip. The pistol is for shooting the cottonmouths that are supposed to fill the river and be thick upon its banks.

In point of fact the river is too cold and swift for cottonmouths, and since I have lived here I have seen only one. He was idling in a warm, shallow pool at the side of the creek that runs along my southern property line. I stopped to look at him from a safe distance. Heavy-bodied and dark, confident and self-assured, he watched me in his turn and did not retreat as a copperhead would have. Instead, he coiled and raised his head, in a defensive posture, ready to strike if I were to advance. He opened his mouth wide, and I could see the white, cottony-looking interior that gives the snake his common name. After he was sure I was not going to come closer, he dropped into the water slowly and with dignity and swam away from me to the bank, where he disappeared under the safe cover of the branches hanging over the pool.

This species of cottonmouth are called *Agkistrodon piscivorus leucostoma,* or white-mouthed *agkistrodon* who eats fish. They are often found in warmish water, and are primarily fish-eaters, but they also feed on other snakes, rodents, frogs and lizards.

The cottonmouth I saw was evidently an old one, for he was big, probably nearly four feet in length; only a slight hint of his cross-banding was visible. Young cottonmouths are lighter, more patterned, and newborn cottonmouths, like copperheads, have yellow-tipped tails. Both cottonmouths and copperheads belong to the evolutionarily advanced group of snakes that do not lay eggs. The young are retained within the mother's body, protected by a saclike membrane, until they are born. Like other snakes, they shed their skins as they grow.

A treasured possession of mine is one of those snakeskins, fragile but perfect, with even the eye scales intact. It is always startling to me to notice people shudder when they see it. There is enough psychomythology about humankind's aversion to snakes to reach from here to Muncie, Indiana, some of it entertaining, much of it contradictory. Whatever the reason, many people are irrationally afraid of snakes, and this makes for poor observation. It is hard to tell what a snake is up to if you are running away from it or killing it. This may account for the preponderance of folklore over natural history in conversations about snakes. It may be why Ozarkers have told me that the hognose snake is poisonous, that snakes go blind in August, that the hills are full of the dread hoop snake who holds his tail in his mouth and comes roaring down hillsides after folks to attack them with the horn on his tail—a horn so deadly that if it gets stuck in a tree, the tree will die within a few days.

My favorite folk story about snakes is the one about copperheads, who are said to spit out their venom on a flat rock before taking a drink of water and then, having drunk, suck it back up into their fangs. I always liked it because I thought it was a grand

Ozark stretcher. But then I found the exact same story in a Physiologus, a medieval bestiary. It is a snake story at least eight hundred years old, perhaps more. So it turns out that the yarn is a piece of natural history after all. It is just that it has to do with the nature of the human mind, not nature of the snakish kind.

There is a magnificent dappled brown and gold house spider changing her skin today in a corner up above the wood stove. A spider grows by molting its skin, which doubles as a skeleton. My spider spent yesterday quietly in her corner, getting it all together and feeling a bit uncomfortable, I suppose, for I read that spiders raise their blood pressure in molt. It has taken her the best part of the morning to crawl out of the old skin, and now she is hanging beside it, resting from the effort, which must have been considerable. Her old skin is beginning to shrivel. It looks wispy and impossibly small for this fine new spider to have worn. She is big, more than half an inch, but not as large as a similar house spider I have seen in the kitchen, so she probably has more molts to go before she reaches her full size.

Molting is one answer to the problem posed by growth—not mine to be sure, but no less correct an answer for all that. Biologists like to emphasize that growth from the inside out is one of the characteristics that separates things that are alive from those that are not. Crystals, which are not alive, grow, but they do so by accretion, simply adding new material to what is already there.

Human beings and other mammals, who hang their soft body parts onto and outside a skeleton, never have to face the growth problems of creatures such as insects—grasshoppers, or the honeybees out in my beehives—or spiders (which are not insects at all, but arachnids). Growth *does* present certain difficulties, but they are different ones.

Spiders, grasshoppers and honeybees—or lobsters, for that matter—wear their skeletons on the outside, and so when they grow too tight they must find some means to shed them. Many insects follow molts with metamorphosis, a complete and radical dissolution of the old body form and rearrangement into a new one. But spiders just step out of their skeletons, doing so anywhere from two to twenty times before they are grown, but keeping the same relative form. Baby spiders, unlike many baby insects, look like adults, but smaller.

One night when I turned on a light in my cabin I found a mother wolf spider with a back covered with babies. Most spiders don't have much interest in their young, but wolf spiders carry them around wherever they go. When the spiderlings emerge from their egg sac, they crawl up on their mother's back and cling to her. The ones I watched, tiny, delicate, perfect miniatures, were an unruly lot and appeared to be causing her no end of trouble. They crawled over her eyes and she had to brush them away. They jostled one another, and several tumbled off onto the floor and then scurried to climb up her legs and return to the security of her back.

The wolf spider and the house spider are both big as spiders go; an outside skeleton imposes mechanical limits on the size of a body that can function nicely. But other, even larger spiders are common in my garden: the black and yellow argiopes, which spin distinctive webs that look as if they had zippers in them. They are brilliant, glossy, stylish spiders, and out there in the garden they trap grasshoppers that eat my tomatoes. They spin winding sheets around the insects and store them until they are needed, like thrifty housewives stocking the larder. It is a way of making a living of which I thoroughly approve, the sort of thing that makes us label the spider as beneficial while condemning the grasshopper as harmful.

I'll admit that I wasn't nearly so pleased one day when I discovered a black and yellow argiope who had spun her web in front of one of my beehives and had stocked her particular larder with tidily wrapped honeybees who had flown directly into it

on their way home, heavily laden with their loads of nectar. I destroyed her web and moved her over to a bush where I hoped she would find something to eat that pleased her equally and me rather more.

Web-weaving spiders don't see very well by our standards; they are so nearsighted that when the males come courting they pluck the strands of the female's web to announce their arrival in order not to be taken for tasty morsels. So I doubt that the beehive argiope was able to see me when I moved her. In a way this was a pity, because for all our differences we share something important. We are both beekeepers; both of us make a living from the bees. My way, compared to hers, seems excessively

Byzantine. I cosset the bees all year long, take away their extra honey, process it, bottle it, truck it to New York to sell to Bloomingdale's, and then use the check to buy the things I need. She simply eats bees.

We are both animate bundles of the chemicals common to all living things: carbon, hydrogen, oxygen, nitrogen, sulfur and phosphorus. Both of us have been presented with a set of problems posed by our chemistry and quickness, among them how to grow and how to make a living. Those are big questions, and as is often the case with Big Questions, we have come up with different answers—answers that in turn are still different from those of the honeybee, who is a similar chemical bundle and upon whom we both depend for a living. The honeybee's solutions have more to do with metamorphosis and the nectar of flowers, and those answers are good ones, too.

Living in a world where the answers to questions can be so many and so good is what gets me out of bed and into my boots every morning.

Last Sunday we held the South Central Missouri Beekeepers Annual Pig Roast. We have it on my place each year in July, because I have space enough in back under the big oak trees for the seventy people who usually come, and also because I am the only one in the group who lives on a river, and they like to have a place to swim. I ask the commander of the VFW post ahead of time, and he always gives permission for the beekeepers and their families to use the beach down there on that Sunday.

One of our members is both a pig farmer and a beekeeper, and it is he who donates the pig. He comes in the morning to get the fire started. He arranges a circle of stones and concrete blocks out in the open away from the trees, and builds up a big fire of oak and hickory logs from my wood pile. While I helped him with it last Sunday, he grumbled to me about all the charcoal factories that are springing up around here. They are bringing in money for our woodcutters, but he is an Ozarker born and bred and he said he didn't like them turning our forests into charcoal briquettes so that people who live in fancy suburbs all over the country can cook on an open fire.

By the time the fire had burned down to a proper pig-roasting stage, some of the other beekeepers and their families had arrived. It was still a few degrees below the hundred predicted by the weatherman, and those who had driven out from town said the slight breeze up here on my hilltop made it seem a bit cooler.

Early in the morning I had set up two sawhorses under the oak trees and put a four-by-eight slab of plywood across them. I threw a bedspread over the top as a tablecloth, and stood back to look. It seemed inadequate preparation for a large party, and the table seemed a little bare, so I walked out into the field and picked an armful of Queen Anne's lace, black-eyed Susans, daisies and butterfly weed, and put them in a bottle of water as a centerpiece.

The beekeepers arrived and their wives began to unpack their baskets. They filled my refrigerator, and put out on the table the food that would not be spoiled by the heat. There were casseroles,

salads made fresh from the garden, a basket of newly picked peaches from a home orchard. One wife had been butchering and freezing her flock of chickens the week before, and had saved a couple to fry for us. We all agreed that chickens which are allowed to run and forage some of their own food always taste better than the kind you buy from the grocery store. There were homemade pickles and relishes, lemonade and iced tea in coolers, pies and cakes that had just been taken from the oven on this hot day. I am not much of a cook, so my contribution is always a washtub full of cold beer and ice. There are some teetotalers among the beekeepers, and they had objected to the beer the first year we had the pig roast, but they come back each time so I guess they worked out some kind of accommodation with their principles.

An interest in bees brings the group together and some of us are friends for other reasons too, but it is a mixed gathering, ranging not only from teetotalers to beer drinkers, but various in other ways as well. Two men are in their eighties and have kept bees all their lives, but there are also young people who have just discovered the pleasures of beekeeping. There are back-to-the-landers with shoulder-length hair and earnest manners, long-time Ozark farmers, the town banker, some schoolteachers and several carpenters. There are retired couples who now have time for a few hives of bees. I am the only commercial beekeeper, the only one who makes my living from it and has a lot of hives. Most of them have a hive or two; a few have ten or twenty. Most of them are men. Nearly all of the women are wives who take an interest in what their husbands like to do.

The children were eager to go swimming as soon as they arrived, so their parents took them down to the river after they had unpacked the food. Everyone had brought folding lawn chairs, and we set them up in a big circle under the oak trees near the table. Those who didn't go swimming sat gossiping and talking bees. Some people came back from the river, cool and wet, and others went down, while some of us stayed behind to watch the fire and turn the pig meat. A few of the new beekeepers had never seen my honey house and were curious about it, so I took them out there, demonstrated the machinery and showed them how everything worked.

By the middle of the afternoon the pig was done and we were all hungry. The women brought the rest of the food out from the refrigerator and arranged it on the table, and we ate as much as we could and then sat in the chairs and talked while the children played. By sunset the temperature had dropped a little, and the wives cleaned up the leftover food. Some of the men carried the plywood and sawhorses back to the barn for me. Families with young children began to leave, but others stayed on.

We sat under the oak trees in the dusk and watched the fireflies rising up from the grass, wondering why you never saw them in the daytime. We talked about how the price of honey was falling and the cost of bee equipment was rising. In the distance the whippoorwills started calling, and I lost the thread of the talk listening for the odd call among them that I have been hearing lately. I think that it may be a chuck-will's-widow, but I don't know birdsong well enough to be sure, so I interrupted the conversation to ask if any of the others knew. As it turned out, there were none among the group who knew bird calls, but everyone stopped to listen and could tell the difference in the call when I pointed it out. We also could hear the katydids and cicadas all around us, and we talked about them for a while and how they made their sounds with their bodies. Then we talked about how poorly the government price-support program was working for small beekeepers.

It was late and cool when my guests decided it was time to go home. A gibbous moon lit the way to their pickups, and there were fireflies everywhere. We agreed that it had been one of our better pig roasts. One little boy had cried when my rooster chased him, but other than that everyone had had a good time.

Last winter was extraordinarily mild, and as a result the chiggers are abundant this summer. Ozark humor is an understated sort of thing, and folks here are asking one another if they happened to attend one of the funerals held for the five chiggers that died during the cold weather.

Once the timber has been cut from these hills, the thin soil will barely support cattle or hog farming and crops won't grow here, so many of the clearings and old pastures go to scrubby second growth and blackberries. That kind of cover and the hot steamy summers make it a prime place for chiggers.

Yesterday a friend of mine from the city was here and went out to pick blackberries for a cobbler. By evening, he was covered with red, ferociously itching chigger bites. They will itch for weeks, although I did not have the heart to tell him so. I was sympathetic because for the first few years I lived here I, too, spent my summers scratching continuously, often in socially unacceptable places. Like many people, after a couple of years I gained a tolerance to them, and so I believe I have now earned the right to take a longer view.

Bad as chiggers are, they have had a worse press than they

deserve because their name so closely resembles that of another pesty being, the chigoe, which is found only in the deep South. Chigoes are insects, tropical fleas that burrow under the skin to lay their eggs; since both chigoes and chiggers are tiny and their names sound so similar, people often confuse them and think that chiggers burrow too. They do not, and they are not insects.

Chiggers are mites and, like their spiderish kin, have unsegmented bodies and eight legs when they grow up. They belong to the class Arachnida, not Insecta. There are more than seven hundred species of chiggers throughout the world, and of these fewer than fifty feed on humans. In the Orient one variety spreads scrub typhus, but on this continent the worst that they can do

is make us itchy. One of our most common species is *Trombicula alfreddugesi,* a mouthful because it is named for Alfred Dugès, who studied them around the turn of the century. Dugès was a celebrated mite man, but my own favorite chigger expert is James

M. Brennan, a government entomologist who once took more than 4,000 chiggers from a single woodchuck in the Bitterroot valley of Montana. Counting out 4,000 tiny chiggers would be an heroic task, and so I was pleased when I read that scientist Brennan had a whole genus named for him.

From a human standpoint, one of the most significant facts about chiggers is that they are so small we cannot see them and seldom realize they have been feeding on us until we begin to itch like my friend the blackberry picker.

Adult chiggers, it is true, are quite visible to the naked eye, attaining a whopping one tenth of an inch. Even I, with my middle-aged eyes, sometimes see them. They are bright orangy red, the color of butterfly weed (or chigger plant, as it is called in the Ozarks). When looked at through a hand lens, the mites can be seen to be covered with feathery hairs, and are rather handsome. However, as they feed on plant and animal debris and an occasional tiny insect egg, our paths seldom cross. But they do lay the minute eggs from which larval chiggers hatch, and it is the larvae that we human beings mind so much.

The eggs, only a hundred microns big, are laid in the soil. Instead of hatching directly, they break open into a second egg, called a deutovum. This second egg hatches the orange larva, a larva so small that it would take a hundred and twenty-five of them, lined up snout to rear, to come up to the inch mark. These larvae must find a meal before they can metamorphose to their next nymphal stage. They crawl up blades of grass, brambles or bushes to find a suitable vertebrate host on which to feed. Their preference would be a lizard, turtle, bird or even a woodchuck; from a chigger's standpoint, a human is a poor host, but should the mite end up on one, it starts to climb to find a protected spot on which to feed. For this reason chiggers most often chose places where clothing fits tightly, and our itches are usually clustered around the ankles, crotch, waist and armpits. Some Ozarkers swear that the best way to avoid getting "chigger bit" is to conduct one's outdoor activities stark naked. Presumably the chiggers would then wander up one side of the body and down the other, discouraged by finding no suitable spot on which to feed. I have never nerved myself to plunge into a blackberry thicket naked, so I cannot report if this is true.

Once a larval chigger has found a good location, he settles down for a feed. Technically speaking, chiggers do not bite at all; their mouth parts are too delicate. Instead, they inject a digestive enzyme into the skin which dissolves a bit of flesh; the chigger then sucks it up. Chiggers are not interested in blood, but feed on liquefied skin and lymph. If left undisturbed, the chigger stays in the same place, using the enzyme to make a small well or tube called a stylostome. If the larva can stay on his host long enough to have a full feed—about three days—he drops off when engorged and crawls to a protected place on the ground, where he rests and prepares to undergo transformation. Inside, his body parts melt down and reconstitute themselves into a legless, pupa-like protonymph, which in turn changes into an active deutonymph, a predator like the adult he eventually will become. The change from nymph to adult is also in two stages. Once again the chigger settles quietly, while within tissue dissolves and re-forms into a legless, pupa-like tritonymph or pre-adult, which does not move, but from which emerges the adult, the bright-red, velvety, sexually mature chigger.

All of which is most elegant and complex, but easily the most curious thing about chiggers from a human standpoint is our own reaction to them. When a chigger commences to feed on us, our bodies overreact by setting off a full battalion of allergy alarm bells, making us itch and scratch. It is actually an overreaction, because it serves no purpose either for us or for the chigger. Those of us who have happily developed a tolerance to chiggers serve as calm, nonscratching chigger hosts. And the chigger is not served by provoking that allergy either. The allergic reaction caused by the venom of the honeybee sting preserves the bee colony and its food stores, but when a chigger provokes an allergic reaction, we scratch it off before it has fed its fill and usually kill it in the process.

Mulling over this curiosity, G. W. Krantz, an eminent acarologist, says, ". . . the intense itching reaction experienced by man . . . reflects a lack of host adaptation." In other words, it is all a Terrible Mistake.

This is one of those biological puzzles that I find cheering—untidy, unresolved, a reminder that the results are not yet all in, that we do not have the final forms, nor all the answers. We are still in process, chiggers, humans and the rest. There are probably better answers somewhere on down the pike.

I keep twenty hives of bees here in my home beeyard, but most of my hives are scattered in outyards across the Ozarks, where I can find the thickest stands of wild blackberries and other good things for bees. I always have a waiting list of farmers who would like the bees on their land, for the clover in their pastures is more abundant when the bees are there to pollinate it.

One of the farmers, a third-generation Ozarker and a dairy-man with a lively interest in bees, came over today for a look at what my neighbors call my honey factory. My honey house contains a shiny array of stainless-steel tanks with clear plastic tubing connecting them, a power uncapper for slicing open honeycomb, an extractor for spinning honey out of the comb, and a lot of machinery and equipment that whirs, thumps, hums and looks very special. The dairyman, shrewd in mountain ways, looked it all over carefully and then observed, "Well . . . ll . . . ll, wouldn't say for sure now, but it looks like a still to me."

There have been droughty years and cold wet ones when flowers refused to bloom and I would have been better off with a still back up here on my mountain top, but the weather this past year was perfect from a bee's standpoint, and this August I ran 33,000 pounds of honey through my factory. This was nearly twice the normal crop, and everything was overloaded, starting with me. Neither I nor my equipment is set up to handle this sort of harvest, even with extra help.

I always need to hire someone, a strong young man who is not afraid of being stung, to help me harvest the honey from the hives.

The honey I take is the surplus that the bees will not need for the winter; they store it above their hives in wooden boxes called supers. To take it from them, I stand behind each hive with a gasoline-powered machine called a beeblower and blow the bees out of the supers with a jet of air. Meanwhile, the strong young man carries the supers, which weigh about sixty pounds each, and stacks them on pallets in the truck. There may be thirty to fifty supers in every outyard, and we have only about half an hour to get them off the hives, stacked and covered before the bees get really cross about what we are doing. The season to take the honey in this part of the country is summer's end, when the temperature is often above ninety-five degrees. The nature of the work and the temper of the bees require that we wear protective clothing while doing the job: a full set of coveralls, a zippered bee veil and leather gloves. Even a very strong young man works up a considerable sweat wrapped in a bee suit in hot weather hustling sixty-pound supers—being harassed by angry bees at the same time.

This year my helper has been Ky, my nephew, who wanted to learn something about bees and beekeeping. He is a sweet, gentle, cooperative giant of a young man who, because of a series of physical problems, lacks confidence in his own ability to get on in the world.

As soon as he arrived, I set about to desensitize him to bee stings. The first day, I put a piece of ice on his arm to numb it; then, holding the bee carefully by her head, I placed her abdomen on the numbed spot and let her sting him there. A bee's stinger is barbed and stays in the flesh, pulling loose from her body as she struggles to free herself. Lacking her stinger, the bee will live only a short time. The bulbous poison sac at the top of the stinger continues to pulsate after the bee has left, its muscles pumping the venom and forcing the barbed stinger deeper into the flesh.

I wanted Ky to have only a partial dose of venom that first day, so after a minute I scraped the stinger out with my fingernail and watched his reaction closely. A few people—about one percent of the population—are seriously sensitive to bee venom. Each sting they receive can cause a more severe reaction than the one before, reactions ranging from hives, difficulty in breathing and accelerated heartbeat, to choking, anaphylactic shock and death. Ky had been stung a few times in his life and didn't think he was seriously allergic, but I wanted to make sure.

The spot where the stinger went in grew red and began to swell. This was a normal reaction, and so was the itchiness that Ky felt the next day. That time I let a bee sting him again, repeating the procedure, but leaving the stinger in his arm a full ten minutes, until the venom sac was emptied. Again the spot was red, swollen and itchy, but had disappeared the next day. Thereafter Ky decided that he didn't need the ice cube any more, and began holding the bee himself to administer his own stings. I kept him at one sting a day until he had no redness or swelling from the full sting, and then had him increase to two stings daily. Again the greater amount of venom caused redness and swelling, but soon his body could tolerate them without an allergic reaction. I gradually had him build up to ten full stings a day with no reaction.

To encourage Ky, I had told him that what he was doing

might help protect him from the arthritis that runs in our family. Beekeepers generally believe that getting stung by bees is a healthy thing, and that bee venom alleviates the symptoms of arthritis. When I first began keeping bees, I supposed this to be just another one of the old wives' tales that make beekeeping such an entertaining occupation, but after my hands were stung the pain in my fingers disappeared and I too became a believer. Ky was polite, amused and skeptical of what I told him, but he welcomed my taking a few companionable stings on my knuckles along with him.

In desensitizing Ky to bee venom, I had simply been interested in building up his tolerance to stings so that he could be an effective helper when we took the honey from the hives, for I knew that he would be stung frequently. But I discovered that there had been a secondary effect on Ky that was more important: he was enormously pleased with himself for having passed through what he evidently regarded as a rite of initiation. He was proud and delighted in telling other people about the whole process. He was now one tough guy.

I hoped he was prepared well enough for our first day of work. I have had enough strong young men work for me to know what would happen the first day: he would be stung royally.

Some beekeepers insist that bees know their keeper—that they won't sting that person, but *will* sting a stranger. This is nonsense, for summertime bees live only six weeks and I often open a particular hive less frequently than that, so I am usually a stranger to my bees; yet I am seldom stung. Others say that bees can sense fear or nervousness. I don't know if this is true or not, but I do know that bees' eyes are constructed in such a way that they can detect discontinuities and movement very well and stationary objects less well. This means that a person near their hives who moves with rapid, jerky motions attracts their attention and will more often be blamed by the bees when their hives are being meddled with than will the person whose motions are calm and easy. It has been my experience that the strong young man I hire for the honey harvest is always stung unmercifully for the first few days while he is new to the process and a bit tense. Then he learns to become easier with the bees and settles down to his job. As he gains confidence and assurance, the bees calm down too, and by the end of the harvest he usually is only stung a few times a day.

I knew that Ky very much wanted to do a good job with me that initial day working in the outyards. I had explained the procedures we would follow in taking the honey from the hives, but of course they were new to him and he was anxious. The bees from the first hive I opened flung themselves on him. Most of the stingers could not penetrate his bee suit, but in the act of stinging a bee leaves a chemical trace that marks the person stung as an enemy, a chemical sign other bees can read easily. This sign was read by the bees in each new hive I opened, and soon Ky's bee suit began to look like a pincushion, bristling with stingers. In addition, the temperature was starting to climb and Ky was sweating. Honey oozing from combs broken between the supers was running down the front of his bee suit when he carried them to the truck. Honey and sweat made the suit cling to him, so that the stingers of angry bees could penetrate the suit and he could feel the prick of each one as it entered his skin. Hundreds of bees were assaulting him and finally drove him out of the beeyard, chasing him several hundred yards before they gave up the attack. There was little I could do to help him but try to complete the job quickly, so I took the supers off the next few hives myself, carried them to the truck and loaded them. Bravely, Ky returned to finish the last few hives. We tied down the load and drove away. His face was red with exertion when he unzipped his bee veil. He didn't have much to say as we drove to the next yard, but sat beside me gulping down ice water from the thermos bottle.

At the second yard the bees didn't bother Ky as we set up the equipment. I hoped that much of the chemical marker the bees had left on him had evaporated, but as soon as I began to open the hives they were after him again. Soon a cloud of angry bees enveloped him, accompanying him to the truck and back. Because of the terrain, the truck had to be parked at an odd angle and Ky had to bend from the hips as he loaded it, stretching the fabric of the bee suit taut across the entire length of his back and rear, allowing the bees to sting through it easily. We couldn't talk over the noise of the beeblower's engine, but I was worried about how he was taking hundreds more stings. I was removing the bees from the supers as quickly as I could, but the yard was a good one and there were a lot of supers there.

In about an hour's time Ky carried and stacked what we later weighed in as a load of 2500 pounds. The temperature must have been nearly a hundred degrees. After he had stacked the last super, I drove the truck away from the hives and we tied down the load. Ky's long hair was plastered to his face and I couldn't see the expression on it, but I knew he had been pushed to his limits and I was concerned about him. He tried to brush some of the stingers out of the seat of his bee suit before he sat down next to me in the truck in an uncommonly gingerly way. Unzipping his bee veil, he tossed it aside, pushed the hair back from his sweaty face, reached for the thermos bottle, gave me a sunny and triumphant grin and said, "If I ever get arthritis of the ass, I'll know all that stuff you've been telling me is a lot of baloney."

Snakes again. Black rat snakes this time. I can't tell one five-foot black rat snake from another, so I don't know if the one that has been showing up in the chicken coop each and every Friday all summer long is the same individual or not, but I rather think he is. My theory is that a week is the time it takes him to digest his meal of mice and an occasional egg. This is theory only, for none of my books tell me when mealtime is for five-foot black rat snakes; it is something I must ask my herpetologist friend the next time I see him.

Black rat snakes are some of the largest common snakes found around here. I estimate the one in the chicken coop to be five feet, but they can grow six feet or more. They are shiny black as adults, but patterned strikingly with brownish and blackish markings when they are young. The vaguest hint of blotchings can sometimes be seen on adults, and this gives them their scientific name, *Elaphe obsoleta obsoleta*. *Elaphe* allies them with their kind, the other rat and corn snakes, and *obsoleta* is a term used in biology to mean indistinct. Their common name hints at their diet.

I have named this one Friday.

Far be it from me to wish a mouse ill, but the mice were rather out of hand in the chicken coop this spring, eating so much of the chopped corn put out for the chickens that I was keeping more mice than chickens. The trouble was that I was fresh out of cats, my pair of barn cats having died within months of each other last winter after fifteen and seventeen years of full and inscrutable lives. Late in the spring, I adopted a kitten, but he had some growing to do before he became a mouser; in the meantime, the mice, unchecked, multiplied rapidly until they became a bold nuisance to me and the chickens.

So I was pleased when I saw the first black rat snake in the spring. There were mice in the barn, mice in the chicken coop, and soon there were black rat snakes of all sizes everywhere. One of the reasons I think Friday must be the same snake is that he has grown self-assured in his sense of possession of the chicken coop, where he soon had the mouse population reduced to tolerable levels. His species can be fierce and will bite if attacked, but Friday seems to understand that I do not intend to hurt him and he ignores me. The day that I found him coiled up in a nest, the

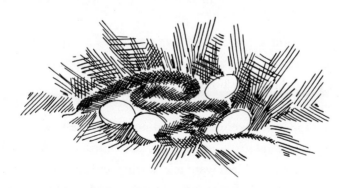

three eggs he had swallowed clearly apparent in his midsection, he looked at me calmly; he was too lumpy to slither away quickly. Last Friday, when I went out to gather the eggs, he was in the coop again. The day was a hot one, and the two-inch-wide circle of water at the base of the chicken-watering fountain had enticed him to try a bit of a bath. He looked me square in the eye as I stood laughing at him. No supposed serpentine dignity could keep him from being anything but ridiculous as he tried to loop and jam his entire five-foot length into the small circle of available water.

Black rat snakes also feed on birds, and in deference to their tastes I brood the dozen pullets I buy each spring in the cabin. I keep them under an electric light in a refrigerator carton near the wood stove. Their downy softness is a delight for a week or so, but they grow gawky rapidly, and stupidly peck one another if they do not have enough space. Their sawdust litter needs constant replenishing; they are untidy with their feed and water, and I soon grow weary of them as roommates. One spring I put them out too soon and the next morning found a dead pullet, too big for a black rat snake to swallow, but small enough for it to kill. They kill animals that size by constriction, and the snake's spiral grip was clearly printed on the pullet's strangely elongated corpse. Now I keep the pullets in the cabin until they are too big to be a snake's prey.

Another time I was able to save a pair of baby phoebes from a black rat snake. The parent birds had built their nest just under the eaves of the honey house, and I had been watching them off and on all spring through the window. The two eggs had hatched, and there were two fledglings in the nest when I was working in the honey house one day and heard a terrible ruckus outside. The parent birds were in a nearby persimmon tree crying out in distress. A black rat snake, like the good climber his breed is, had slithered up the side of the honey house and was looped around the nest calmly swallowing the two baby birds. I ran outside, grabbed the snake by the tail and shook him hard. The baby birds dropped from his mouth, wet but undigested. I threw the snake as far as I could, scooped up the babies and put them back in the nest. The parent birds remained in a state of ineffectual confusion all day, alternately repelled by and drawn to their offspring. At nightfall they finally returned, and the pair of young phoebes lived to fly from the nest on their own.

And there we are, with my meddling, back to the human responsible for putting a flock of chickens in prime mouse habitat, setting the process in motion in the first place. I like to think of it as a circle. If I take one step out of the center, I find myself a part of that circle—a circle made of chickens, chopped corn, mice, snakes, phoebes, me, and back to the chickens again, a tidy diagram that only hints at the complexity of the whole. For each of us is a part of other figures, too, the resulting interconnecting whole faceted, weblike, subtle, flexible, fragile. As a human being I am a great meddler; I fiddle, alter, modify. This is neither good nor bad, merely human, in the same way that the snake who eats mice and phoebes is merely serpentish. But being human I have the kind of mind which can recognize that when I fiddle and twitch any part of the circle there are reverberations throughout the whole.

My honey house is a lean-to on the south side of the barn, with windows on all free sides. Fastened to the east and west windows there are bee escapes—twists of wire that the bees can leave by but not come through—and out of these go the bees who were reluctant to leave the honey supers in the bee yards and have returned to the honey house with Ky and me. We stack the supers in the honey house while they are awaiting processing, and gradually the leftover bees crawl up through the frames. They are dispirited when they get to the top because they do not know where they are, but bees instinctively fly toward light, and so, gradually, over a period of days, they discover the bee escapes on the eastern window during the morning when the light comes from that direction and the western one in the afternoon when the sun shines on that side of the honey house. Outside they are as confused as ever, but above the bee escapes I keep half-sized hive boxes with a few frames of honeycomb in them, and the bees enter them gratefully. After the honey harvest is done, I take the bees who have gathered in those two boxes and shake them into some of the hives I keep back by the woodlot.

Doing this is putting bee biology to work in a textbook way to clear the honey house of bees. But this year I had some bees who hadn't read the textbooks, and their behavior puzzled me.

For the first week of honey processing, the loose bees inside the honey house were finding their way out at the bee escapes, and the usual hordes of bees were on the outside of the screens, smelling the honey inside and yearning to get in. Sometimes the ones on the outside and those on the inside would feed one another, poking their long tongues through the wire to make contact. In addition to feeding, I knew that they were telling one another things, for chemical communication in the process of food exchange is one of the regulators of bee behavior inside the hive. One day as I watched them I idly wondered what information they could possibly have about a world as artificial to them as the honey house.

During the second week, a small cluster of bees stubbornly refused to leave one of the southern windows. The cluster grew larger each day as more and more bees gathered. Bees are intensely social, and the more that join together, the better their morale becomes. They even built a few inches of honeycomb on that southern window with which to cheer themselves. The outside bees began coming more often to the spot on the screen where the obstinate bees inside were clustered. Feeding between the two groups was constant. Having bees on the inside of the honey house is cruel and untidy, and I tried a number of tricks to get them to move, cutting down their comb and dispersing them, trying to entice them onto a frame of comb that I could shake outside. But they defeated me and returned to build back the comb, several thousand strong by this time. In twelve years of keeping bees I have learned that I can't make bees do what they don't want to do, so I gave up and watched the cluster of bees grow. By the third week of bringing back new supers from the outyards, it had grown to the size of a small swarm.

We spent the following week processing. One day I was alone in the honey house at sunset emptying the 1200-pound storage tanks, draining the honey out and putting it into sixty-pound plastic buckets. The sun was shining low into the western window, and some newcomer bees were finding the bee escape there just as they were supposed to do, so I did not turn on the overhead lights lest I confuse them. The big cluster of bees still hung perversely on the southern window. I stopped my work to look at them, for there were more than the usual number of bees opposite them outside the screen. They were agitated, their ex-

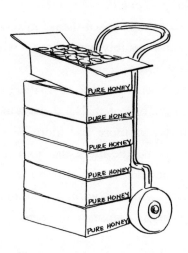

citement palpable. Suddenly, as if on cue, most of the bees in the cluster started walking away from the southern window, leaving just a few behind on the solacing bit of honeycomb, walking across the ceiling purposefully, not toward the western window and the light, but the opposite way, into the shadows. They walked in long lines across the darkened ceiling toward the eastern window, where they massed at the bee escape, piling on top of one another. They jostled and pushed their way through the bee escape, and within fifteen minutes those thousands of bees were outside.

What had happened? What made the bees leave their bit of comb, the only happiness they had been able to contrive in the honey house? Why did a few stay behind? Why had they walked away from the light? How did they know that there was a way out in the shadows? Why did they walk in lines? What signal made them all move together?

I have no answers to those questions, but I presume that the constant communication with the bees outside the screen had something to do with their behavior, and this suggests a greater complexity of information conveyed and understood than I feel comfortable about. An entomologist I asked had no answers, either.

It was less of a puzzle to a fellow beekeeper to whom I talked. "Those bees on the outside just told the bees on the inside how to get out, that's what happened," he said. I said I didn't know. He shook his head. "It's spooky, that's what it is. Just plain spooky." And we both agreed that the longer we keep bees the less we understand them.

I'd been turning these questions over in my mind for some days when I remembered another equally baffling incident of insect direction-finding ability.

I was visiting my cousin Asher, an entomologist, when a neighbor stopped in with a stalk of milkweed on which munched a fine monarch butterfly caterpillar, handsome in his white, yellow and black bands. "When I found this, I thought of you," she said, and we all laughed because in some company that might be a strange thing to say. Asher promptly named the caterpillar Henry, put the stalk of milkweed in a luminous green vase of water to keep it fresh and placed it on a table where we could watch.

We spent the afternoon sitting, talking and idly looking at Henry, who had already eaten several leaves. While we watched, a leaf on which heavy Henry was feeding gave way and he fell with a plop to the table. There were a variety of manmade objects there, but Henry paused only seconds to gather his caterpillar wits, then unerringly crawled toward the green vase and clumsily scaled its glassy, bulging surface to return to his milkweed leaves.

Summer 82

Asher's specialty is moth ear mites, but he also has some knowledge of moths and butterflies; however, when I asked him for an explanation of what we had seen, he was as much at a loss as I was to understand how the caterpillar had found his way in what seemed such a purposeful manner amidst an environment in which his instincts could not have helped him. We talked back and forth about it for a bit, and in the end shook our heads in wonder.

Despite the commonness of their seasonal behavior, we have no real understanding of how birds find their way when they migrate. Theories have been put forward suggesting that birds make use of landmarks recognized by experienced members of a flock, use celestial and astronomical navigation or have a magnetic sense that allows them to detect slight variations in the earth's electromagnetic field, but all of these theories are unsatisfactory in some respects. I recently read that researchers have discovered birds can hear sound traveling thousands of miles at frequencies which pass right through objects, and which emanate from air disturbances around large geographical features such as mountains or oceans. This means that birds migrating across my farm in Missouri may be hearing the Rocky Mountains or the Atlantic Ocean, and taking their bearings from what they hear.

Whatever clues they use, Robert Crawford, a young Florida researcher, said, "It's like they have another dimension. We sit here in our comparatively dull world thinking that we know all and see all."

He was talking about birds, but he might have been speaking of the monarch caterpillar or my honeybees. I believe it was Sir James Jeans, the physicist, who was supposed to have observed that we live in a world that is not only queerer than we think but queerer than we *can* think. I remembered this the afternoon I watched the bees streaming across the ceiling in the honey house, and was grateful that every once in a while I can have a glimpse of just how queer that might be.

AUTUMN

I spent the afternoon today astraddle the ridge of the new barn-loft roof laying down a ridgecap, a course of overlapping shingles that covers the seam where the shingles meet from each slanting roof face. It would have been difficult not to have been happy up there. The fall migration of monarch butterflies, stunning creatures of orange and black, has begun, and numbers of them flew past me while I worked. The sumac along the woodlot edge has started to redden. It was more vivid from up on the roof, and I could see deep into the woodlot as well. The new roof is three feet higher than the old one and, standing up, I could see north beyond the river, over ridge after wooded ridge, farther than I have been able to see before. After a few days of rain, the skies have cleared to the deep blue that reminds me that Missouri is a part of the West.

The old barn-loft roof, a cobbled construction of tin, was held, in some places by wire, to misplaced, undersized roof beams, and was not much of a barrier against wind, rain and snow. In storms parts of it often blew off, and I would gather them up and try to nail them back on the rotting beams. Every rainstorm meant a rearrangement of buckets to keep the leaks from puddling through the loft floor to damage my workshop below. I used the loft for storage, but everything in it had to be draped in plastic to keep it dry.

This past summer I had finally put together enough money to buy materials for framing and a new roof; my son, Brian, and his friend, Liddy, came from Boston, where they live and are finishing graduate degrees in architecture, to build it for me. They had been working together on a construction project all summer, and were tan and fit. They are both of small build, but muscular, well coordinated and strong. The first day they were here we went down to the lumberyard together, and on being introduced one of my friends there slipped his fingers across Liddy's palm as he shook hands. "Well, she's got calluses at least," he said grudgingly. He had not believed in this lady carpenter I had been telling him about.

Liddy and Brian tore down the old roof and raised the loft walls three feet for headroom. They framed up the new roof, put it on and shingled it. In the additional three feet of wall space, they set out, at a dramatic angle, all the old windows that I have accumulated at farm sales over the years and had painted with bright tractor paint, International Harvester red, John Deere green and yellow, Ford blue, and Allis-Chalmers orange. Now the space inside the loft is so cheerful, so light, so airy, that I want to live up there. They extended the ridgebeam from the peak both fore and aft, and with the addition to it of a few pieces of scrap wood cut in curves. Brian gave the roof a head and tail, transforming the barn into a large friendly animal crouching on my hilltop. With carefully bent and glued strips of thin wood, he laminated a graceful arch below the roof peak in front. My neighbors soon heard about the arch and have been dropping by for inspection. They stand for a few minutes looking at it, and

then some of them say "Wow!", but others just nod in approval.
It is now that sort of a barn.

Liddy and Brian work well together. It takes only few words
for them to understand each other's thoughts and intentions. I ran
errands for them to the lumberyard and occasionally helped, but
discovered soon enough that when I offered to either stay and
help or go bake a pie, it was the pie that was more needed. It
took six pies to finish the roof. I had not known that pies were
such an important part of construction.

One day, when the temperature was nearly a hundred degrees,
I talked them down off the roof and we took inner tubes to the
river. We floated a couple of miles, splashed about and made the
acquaintance of several gravel bars and the fish that swam near
them. In the evenings after dinner we sat over coffee and talked
and talked.

When I was pregnant with Brian, I was pleased, curious and
interested, but somehow detached and objective about the baby
I was carrying. I was young, and had no notion of what he would
mean to me. I was anesthetized during his birth, as women were
in those days, and anesthesia brought a dream, a revelation: the
secret to mailing dogs was to make sure that their ears were in
mailing tubes. That done, any dog could be mailed anywhere in
the world. It was an epiphany I struggled to share with the nurses
and the doctor as I slipped back into consciousness. I was irritated
with them because they wouldn't pay attention, and instead
insisted that I look at the new baby. I was cross, loud, babbling
about dog ears and mailing tubes until they forced the baby into
my arms. I looked at him, blond, blue-eyed, beautiful, and was
overwhelmed by the rush of emotion that traveled through my
body. I was seized by motherhood, and unprepared, stunned by
it.

There has always been a part of me that stood aside, watching, commenting, and that piece of me, despite the lingering anesthesia, was in good form. "What is happening to you?" it asked. "Pay attention. This is important."

My baby. *Mine.* I became impossibly stubborn when the nurse suggested it was time to take the yet-unnamed boy back to the nursery. I wanted to hold him always. My baby.

There was a fierceness to the love that was born the instant I saw him that startled and bewildered me. It was uncivilized, crude, unquestioning, unreasoning. I first began to understand it when, several years later, we were on a family camping trip, and during the night were awakened by an old sow bear who had wandered into our campsite with her cub. Her baby had strayed to the other side of our tent. She was frantic, fierce, angry, and would have become dangerous had not the cub waddled back to his mother of his own accord.

Because I am not a sow bear I did allow the nurse to take Brian the evening he was born, but I cried when I let him go. In order to become an adequate mother, I had to learn to keep the old sow bear under control. Sow-bear love is a dark, hairy sort of thing. It wants to hold, to protect; it is all emotion and conservatism. Raising up a man child in the middle of twentieth-century America to be independent, strong, capable and free to use his wit, intellect and abilities required other kinds of love. Keeping the sow bear from making a nuisance of herself may be the hardest thing there is to being a mother. Over the years she snuffled about when he learned to walk and explored the edges of high places, whenever he was unhappy, when he went off to boarding school, when he started driving a car. It was the sow bear who fifteen years ago pushed me into organizing a chapter of a peace group, and to become a draft counselor on the Brown University campus where I was working. It seemed just possible in those days that Brian might turn eighteen with a war still in progress. Even if I had to organize a whole peace movement to keep it from happening, no government was going to be allowed to send my son off to war.

I did learn to live with the old bear. Fortunately, so did Brian; he and I understand one another pretty well.

Liddy and Brian finished the most important parts of the barn before they left, but they ran out of time and the lumberyard had not received some of the materials Brian had ordered, ridgecap shingles among them. Liddy showed me how to use several tools that I had not used before, and laid out the angles for me to cut the remaining window supports. Brian showed me how to cut and lap the shingles for the ridgecap, and reminded me several times how necessary it was to make the roof tight from rain.

When they got ready to leave we all agreed that we had spent a fine time together, that the barn was a thing of beauty, that the pies had been good, that we all loved one another. We hugged and kissed. And there was no need at all for the old sow-bear tears in my eyes.

Brian put his arm around my shoulder one last time.

"Ridgecap," he said, smiling gently down at me. "Remember, no guarantee on the job without a ridgecap." I laughed, nodded, and promised that I would take care of it.

And today, up there in the bright sunshine with the monarch butterflies, I did just that.

I keep detailed records arranged by outyard on all my beehives. In this way I can tell how the hives are coming along and what needs to be done with them, and also how productive the bees are in each yard. Sometimes bees in one place stop making much honey because a nearby farmer has cleared a piece of land of wild nectar-producing flowers and has turned it into pasture. When this happens I move the hives out in the autumn to overwinter here, and set them in what I hope will be a better yard the following spring.

I have had an outyard upriver from me that has been consistently poor. I should have moved the bees long ago, but I kept finding excuses for leaving them there because the old man who owned the farm liked them so much. He has lived alone for many years, and some people say he is a bit tetched. I never found him so; he *was* lonely, however, and liked to talk when I would take him his gallon of honey each autumn. We spoke about bees, for at one time he had kept them. We discussed cows, for he had been a dairy farmer and knew a lot about them and I didn't, so I liked hearing his stories. He had a productive orchard which he never sprayed for fear of hurting the bees who pollinated his fruit, and I always was interested in how it was coming along. Once he told me about his son, who had died twenty-five years ago. He cried when he talked about him. I didn't know what to say. He loved the farm, one of the prettiest along the river. It was all he had left, but it was becoming too much work for him, and for several years he has been trying to sell it.

When I went to take the honey from the bees in August, a SOLD sign was by the edge of the road and a heavy chain lay across the roadway leading down to the farmhouse. Fortunately it was on the ground, so I drove across it, down the road and then into the woods where the bees are. The day I went to take the gallon of honey as rent, the chain was stretched tautly from two new posts and held in place by a padlock. A freshly painted sign said KEEP OUT. I reminded myself to find out who the owners were and to telephone them.

Before I had a chance to do so, the new man phoned me. He sounded irritated. He had been out walking his boundaries, he said, had seen some beehives with my name and telephone number on them, and wanted to know what they were doing there. I explained the arrangement that I'd had with the old man, and that I had a gallon of honey rent to pay if I could get in. He wasn't interested in any honey, and he wanted those bees out of there right away. I told him I had been intending to move them for some time, and would be doing so later this fall, when it got a bit cooler and the bees were easier to move. Mollified, he asked me to telephone him before I came so he could open the roadway; he said he'd also like me to stop by the house because he'd like to have some advice about how to deal with all the wild bees around.

"Wild bees in trees?" I asked. Sometimes my bees do swarm and take up homes in hollow trees.

"No, wild bees hanging all over in those . . . those whatyou-callums . . . cocoons . . . hanging all over on bushes," he said.

His rapid speech suggested that he was not an Ozarker, and if he was talking about hornets, his lack of knowledge suggested he was not a countryman. I asked him if he meant large football-sized, round, gray, papery-looking nests.

"Yes, yes," he said, sounding pleased for the first time. "Those are the cocoons."

"Those aren't really cocoons," I explained. "Those are nests, but bees don't live in them. They're hornets, bald-faced hornets."

"Well, bees, hornets, whatever. I don't want 'em around. Nests either."

This has been an ideal year for bald-faced hornets and they have been unusually numerous, a fact that should make an orchardist happy. But when I tried to explain this to him, he brushed it aside and asked me how to get rid of them. I advised him to do nothing until the weather got colder because the hornets might still be in the nests. I promised to stop and talk with him when I moved the bees, and we ended our conversation on cordial enough terms.

Adult bald-faced hornets can eat only liquids because of the structure of their mouthparts. Their diet, chiefly carbohydrates, is usually flower nectar, but they must have a source of protein for developing larvae, and for this they kill enormous quantities of caterpillars. I often see them in the summer when their brood-rearing is at its peak, flying low over the ground or around tree trunks looking for prey. When they find a caterpillar, they do not sting it but butcher it alive. First the hornet kneads the caterpillar with her mandibles to soften the muscles and other tissues, then cuts it up with her mouth. She swallows the liquid parts herself, and forms the solid parts into pellets which she carries back to the nest. Here nurse hornets take over the bits of flesh to break into still smaller morsels with which to feed the developing brood. Considering that there may be 10,000 hornets in one of the big nests, the requirements of a single colony may serve as a considerable check on destructive caterpillars in an orchard. A number of nests should cheer an orchardist.

Bald-faced hornets, *Vespula maculata* (*vespa* is Latin for wasp and *maculata* means spotted), are big stout insects, three quarters

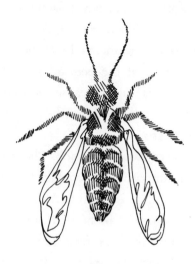

of an inch long, black with whitish yellow markings on the head, thorax and abdomen. A mated female who overwinters on the ground under a bit of leaf litter or stone starts a new colony afresh each spring, chewing up pieces of wood to make the papery material she needs to construct a small, hanging starter nest of closely arranged cells in which she lays the first eggs. The tubular cells are covered with another protective layer of the same gray, papery material. The first eggs develop into workers, sterile females. They enlarge the nest and start killing caterpillars for the next generation. By summer's end, males, who grow from unfertilized eggs and lack stingers, are born, and they fertilize a new generation of females. In the autumn all of them die except for the fertilized females, who abandon the nest for better protection against winter's cold.

Adult worker hornets are extremely protective of their nests and are more aggressive than most wasps. Unlike honeybees, they do not lose their stingers after they have used them and they will sting repeatedly. A few summers ago, bald-faced hornets would sting and chase away anyone who tried to sit under the big oak trees out in back of my cabin. I never found their nest, but I was stung a number of times. Their venom, different from a honeybee's, triggered a reaction in me, and I found the stings painful.

I asked the new man at the beeyard if he had been stung, thinking perhaps that this was the reason for his dislike of hornets as well as honeybees. No, he had not; he just didn't like having bugs and bugs' nests around.

Moving beehives is a two-person job. The shape of two-story beehives make them impossibly unwieldly for one person to lift into a pickup; besides, they are heavy. At Thanksgiving, the son of some friends of mine will be home from college, and he has agreed to help me move the bees from the outyard. I will be glad to bring them home, and next spring will take them to a better yard on a farm where people really want them.

I shall stop in to see the new man whom I talked to. I am curious to meet him, and I shall show him how to make a torch to burn the old empty hornet nests. The café in town has an especially big one nailed up in a corner and I shall tell him about it, but I do not know if he will be interested.

Andy Beagle was out at dawn this morning doing a bit of rabbit work. He is twelve years old now, and stiff in the joints sometimes, but on a misty morning when scents cling to the grasses there is nothing like hunting rabbit after breakfast to keep an old guy in shape.

He likes being called Andy Beagle. He wags his tail when I say to him, "Andy Beagle, Andy Beagle, Andy *Beagle* is his name and he's a *good* dog."

In truth, although I do not mention it to him, he is probably part foxhound. The first summer that Paul and I lived here we found him in the road one day in the company of a sister, the pair of them about the size of rats, and a mother, a full-blooded beagle who was thin and desperate. The three of them were starving and nearly hairless from bad cases of mange. I do not think they had been dumped out, because people here are serious about hounds. Perhaps the mother had been lost one night when her owner had taken her out to run with other hounds, and her pups had been born since then. She looked as though she had been on her own for a long time. When we asked around no one knew of her or her pups. We already had two other dogs, a pair of Irish setters, but we brought home the hound and houndlings and fed them. The female pup was too far gone and she died the first night. But the male, weak as he was, looked up from his first dish of food, eyes shining, and gave me a little toss of his head, a gesture that I came to learn is a beagle heigh sign, and I knew

he would live. The mother, too, regained her health and strength, and with the application of the same mange medicine with which I was treating her son, her coat also grew back. She was grateful for the food, but she was old, wary and had seen a good deal of life. After she ate, she would retreat under the pickup and never presume to become a part of the affairs of this place. A man I knew who bred beagles came over one day and said that she was a good one and he would like her for breeding, so I let him take her.

The male pup, however, had settled in self-assuredly, and won me over by being the only dog I have ever known who knows how to hug. He thrusts the top of his head against my neck under my chin and presses hard against me. I do not know whether other beagles do this, but Andy has done it all his life. I have always named pets after writers, so we named him Andy, after E. B. White. He grew leggier than is usual for a beagle, and his black, white and liver markings and build suggest that his father was a foxhound.

Over the years, with no encouragement from any human, Andy has turned himself into a hunter. Nose to the ground, he follows a trail, baying as long as the scent is present, silent when it is not. He is dedicated and passionate, and very occasionally he catches a rabbit. But not often. Sometimes I wonder whether the old and experienced rabbits about the place take him out for a run in the mornings just for exercise.

The one who more often catches a rabbit is Tazzie, the new dog about the place, whom I acquired some years after the setters died of old age. My niece had taken in Tazzie's mother, a Belgian shepherd, from a man who was moving and wanted to be rid of her. He claimed that she was spayed, but it turned out that she was pregnant and gave birth to seven pups of undetermined fatherhood. My brother Bil who writes things had just returned from Tasmania, where he had been doing a magazine story on Tasmanian devils, and when he saw the litter he singled out my dog and insisted that she had the markings of a Tasmanian devil. The name Tazzie was born. When I took her I said I would keep the name, even though it was not a writer's, on the understanding that it was short for Tasmania, which has a regal sound she may earn in her mature years, but not devil; for there was no hint of deviltry about Tazzie and she showed every sign, even as a pup, of having all a shepherd's qualities: loyalty, lovingness and obedience.

Even while she was just a black fuzz-ball puppy, Tazzie tried to follow Andy everywhere, but the shepherd in her made her a sight, not a scent, hunter, and she found it hard to understand the reasonableness of running along with one's nose to the ground. Eventually, as she got older, she discovered that when Andy started baying and carrying on, it meant that a rabbit was about, and that if she just sat quietly in a central place a rabbit might run right in front of her, and that if she were quick, she might—sometimes—catch one. The rabbit population is not in much more danger from her than it is from Andy, but the two dogs spend hours hunting, each in his own way.

Today I discovered that Black Edith, the new cat, is wilier and even more of an opportunist than Tazzie, and may become the most effective hunter of all. I got Black Edith last spring from a farmer who, upending the two black kittens in the litter, handed me what he said was the female. I had named a previous black cat, a male, Sacheverell, so it seemed only proper to name this one Edith. Soon, however, I noticed that Edith was growing testicles. Any sexually mature male cat I have ever had has always wandered off, and I did not want to lose this one because he has such a fine personality. After spending the night outdoors doing whatever it is cats do out there, he greets me with a pleased-sounding *per–r–r–r–k* each morning, wrapping himself around my ankles in an ecstasy of affection while I make the coffee. When I sit down to drink it, he jumps on my lap, purring and enjoying the morning coffee ritual in his own way as much as I do. In addition, he is intrepid. The dogs would like to chase him, but when they try he rolls over on his back and swipes at their noses and they back off, acting confused. It is true that Tazzie regards

Black Edith as a living bone. I often find her gnawing on him, holding him between her paws, but he regards this as a pleasant display of attention, and when her teeth become too serious, he yowls and rakes her tender nose with a full set of claws until she lets go.

I wanted to keep Edith, so I took him to the animal clinic to be neutered. As the vet's wife was making out his admission card she suggested that I should rename him Eddie. No, Edith it is, I insisted, and Edith it shall stay. However, he has grown up to be an emphatic cat, and it has become more fitting and proper to call him Black Edith.

This morning I was sitting in the brown leather chair listening to the news on the radio, dimly aware of Andy baying some-where in the distance as he pursued the elusive rabbit. Through the three big windows I could see Tazzie sitting calmly on a strategic high spot in the field. My attention, however, was on the news until I saw a black shape bounding toward the barn. It was Black Edith, carrying something nearly as big as himself in his jaws. It was the rabbit. Pausing momentarily to study the twelve-foot post connecting the deck to the barn loft, he leaped to the deck, touching the post at one point only. Although by now staggering, he again leaped, sideways this time, out over space into an as yet unfilled window hole in the new storage loft. He was correct to move quickly, for Tazzie had spotted him and came running. She hurled herself up the loft stairs, but of course was not agile enough to leap through the window opening. Frustrated, she scratched the loft door and whined. Some time went by before Andy, nose to the ground, picking up scent now of both rabbit and cat, arrived. He sniffed at the post Black Edith had bounded from in his leap, looked up, understood what had happened and joined Tazzie on the deck. Both dogs could hear activity inside the loft and glared at the door, outraged that they had been done out of their rabbit by a mere cat.

By this time I had come out, and could hear noises inside the loft that suggested Black Edith, young and inexpert as he is, had not killed the rabbit but was chasing it around the loft, upsetting things as he did so. I put both dogs in the cabin and opened the loft door. Black Edith greeted me with an exasperated *mee-owwww,* hoping, perhaps, to ally me with his cause. The rabbit was hopping about unharmed. I scooped up the cat and took him to the cabin, then returned to the loft to look for the rabbit. I soon found him in a corner behind some boxes, frozen with panic terror. I picked him up and discovered that despite his wildly beating heart he had only the slightest of skin wounds, and would be able to survive in the wild. I walked a quarter of a mile behind the barn, out of sight of the dogs and cat who were watching me intently through the cabin windows. While I walked I stroked the rabbit and felt his heartbeat quiet. He was a young rabbit, also inexperienced, and perhaps he learned something useful this morning. Back in the woods I found a thicket of fallen tree branches and briars, and set him down there. He looked at me for rather a long time and then hopped away deeper into the woods.

Today is my wedding anniversary. I am sad when I think of Paul and remember what expectations we had the day we were married. But those failed expectations are what made me a beekeeper in the Ozarks. I like that a lot, and none of it would have happened without Paul.

My grandfather was a beekeeper in Kalamazoo, Michigan, where I grew up, but everyone's grandfather was a beekeeper, so that doesn't signify. Besides, my grandfather frightened me. He terrified everyone within shouting distance and I tried to stay out of his way so I picked up no love of beekeeping from him.

He was a stern man who always wore three heavy layers of clothing, starting, it was reported, with a woolen union suit. He claimed that all the layers kept heat out during the summer and in during the winter.

Grandpa liked to winter his bees in the basement, and in the autumn he would imperiously order my father to carry them down there single-handed, despite the danger of hernia. Grandpa couldn't help because he was subject to terrible, racking pains caused, the doctor tried to tell him, by the binding of the union suit.

When Paul and I first started keeping bees and reading about beekeeping, I discovered that overwintering bees indoors is an old-fashioned practice now discredited. So much for the grandfatherly example.

My grandmother was a timid, sad-faced woman, worn down with the cares of living with such a man, and trying to manage on the tiny household allowance he doled out to her. She never complained and seemed almost saintly. She outlived him by many years, and after his death regained a measure of spirit. Toward the end of her life, she gathered her grandchildren around her.

"I want you to remember your grandfather always," she said.

We nodded solemnly. She beckoned us closer. "I want you to remember that he was a mean, dirty, stingy old man," she said in a firm voice, and then looked off into the distance, a pleased smile playing over her face.

My other grandmother, Annie, was quite a different grandma, but she didn't have anything to do with bees either. What she had to do with was Success and Men.

She was a correct and regal woman despite the fact that she walked with a slight limp. She had injured her knee from a fall during a bicycle race. Everyone said she had been a demon cycler in her younger days. Back then, in the 1880s, she had borrowed money from a Man banker at three percent interest to go to college. After graduating she taught school, paid back her loan, raced bicycles and, clawing her way past Man players, became a state tennis champion. She was competitive and adored sports. I can remember her hunched over the radio, muttering at the Chicago Cubs, who were a constant source of disappointment. "Just like a bunch of Men," she would say.

She would have sniffed at today's debate over the Equal Rights Amendment, for she believed that she and all her sex were superior to Men, and that mere equality would be a step down for any woman. In fact she would have sided with the anti-ERA forces, if only because of the unisex bathroom issue. I can remember going on Sunday afternoon drives with her and stopping at gas stations with only one restroom. She would go in to use it, but would stalk back, tall, straight, her eyes blazing. "Man pee!!!" she would announce loudly, and not even the greatest need could force her to use that single-sex facility.

Grandma Annie had been married briefly and irrelevantly; her two daughters were the product of immaculate conception. There were only two men who won her approval. "Yes," she would sigh, "we've all been persecuted: Douglas MacArthur, Jesus Christ and me."

I never found out exactly how it was that she had been persecuted, but persecution, she said darkly, accounted for her own lack of Success in Life. It had escaped her, but she was determined it should not escape her grandchildren. None of us showed signs of athletic ability and would never be scouted by the Cubs, so it was hard to know what she wanted for us. We thought about it a lot. Occasionally, instead of saying that someone was a Success, she would speak of him as a Big Person. I knew that she had her heart set on my brother Bil growing to be six feet tall, so I thought that just growing very large might be satisfactory. Once I took our family dog, a brainless sway-backed Great Dane whom Grandma Annie hated, to a pet show, and the dog, for all her faults, was awarded a beautiful blue silk ribbon with gold lettering for the pet with the longest tail. I showed it to Grandma Annie, and ever afterwards she spoke well of the dog. I got the impression that if any of her grandchildren had grown very big and been able to grow tails of suitable length, she would have been proud.

As I grew older, however, I learned that Success meant more than physical stature. We grandchildren were to win foundation grants, a Nobel prize each, and the fourth-grade class presidency. I think she had in mind universal acclaim, esteem, admiration and a certificate suitable for framing.

When I turned three years old, Grandma Annie decided that I should become a concert pianist. My father bought a piano and my mother found a teacher for me, a Roman Catholic nun named Sister Esther who taught music at a convent school. Our family wasn't Catholic, and I had never seen a nun before. I was able to take in stride her long black habit, her cross on a chain and her rosary beads, but I could not help staring at the stiffly starched white collar that covered her neck and encircled her face. It made creases in her cheeks and forehead that I could see when she turned her head. I felt sorry for her and tried to be good.

Sister Esther was a severe, tense woman. She showed me a naked white china doll in a toy crib and a pile of straw nearby. The doll, she told me, was the baby Jesus, and he was cold. If I practiced hard and had a successful lesson, I would be allowed to put one straw in the baby Jesus' crib to help keep him warm. But if I was not successful, I couldn't put a straw in the crib, the baby Jesus would get a chill, and it would *all be my fault!*

I was appalled.

Like many children, I had been born with perfect pitch. Aided by that biological accident, driven by trying to be nice to an oddly dressed woman with creases in her face who knew what Success was, but most particularly horrified that the baby Jesus might catch cold on my account, I quickly learned the scales on the piano. Three lessons went by, and I was able to give the baby Jesus a straw each time.

Sister Esther told my mother that I was a child prodigy. Sister Esther was pleased, for I was her first prodigy. She even smiled a little. Grandma Annie said it was only natural that her grandchild should be a prodigy.

I was terrified. I was sure that those scales had been my limit, and I was right. I was only three and a half years old and, musically speaking, I had peaked. As the lessons went on, Sister Esther grew stranger and more agitated. I couldn't understand what she was trying to tell me about harmony. The baby Jesus never again got a straw. My fingers stumbled when I tried to play exercises. Sister Esther fingered her rosary. I embarrassed her by forgetting my memorized piece at recital time. The baby Jesus suffered, she told me. I was miserable.

The years went by and I was still playing the piano at the level of a precocious three-year-old when Sister Esther had a nervous breakdown. Just before she did so, however, she tensely admitted that she might have made a mistake about me, and that actually I was rather retarded musically.

Grandma Annie recovered quickly. She said, "Well, maybe the child can dance. She has a very long neck." She was always saying that kind of thing. I couldn't understand what the length of my neck had to do with dancing. I felt like a giraffe, and tried to pull my pigtails around my neck so that people wouldn't notice.

Mother found a blowsy retired dancer who taught ballet in her own home and enrolled me as her pupil. Since I was an awkward, gangly child, the ballet teacher sized me up right away. She set me some simple exercises, sat down in a comfortable chair with a flowered slipcover and sought relief from watching me by taking a nip now and then from a dainty flask she kept tucked in the bosom of her leotard. The nips gradually became more frequent over the months until, at last, she told my mother I was hopeless. She took down her ballet-teaching sign and joined Alcoholics Anonymous.

I was very sorry that I was already eight years old and had caused so much pain to the adults, had added my share of persecution to the baby Jesus and was not a Success (although I was grateful I had not harmed Douglas MacArthur). Deep down inside I knew that I was never going to be a Success because of something I had learned in school: all the Successful people were dead. They had taught us about George Washington. He was wise, calm, patriotic and truthful. He was a Success. He was dead. They taught us about Alexander the Great. He found a knot that couldn't be untied (the kind that you get in shoelaces sometimes, I surmised), and he cut it in two. Children were not allowed to cut the knots in their shoelaces, but when Alexander did so it showed that he was an innovative thinker. He was a Success. He was dead.

They taught us about Robert-the-Bruce. He wanted very much to be king, but failed a number of times. Once he failed so badly that he was caught and put in a dark dungeon. While he was in the dungeon, he used his time by watching a spider spinning a web in a corner. Over and over again the spider tried to attach her web where she wanted it but kept failing, until at last, with a huge effort, she managed to get it just right. This encouraged Robert-the-Bruce so much that he went on trying to be king while he was still in prison, and after he got out too. Teachers in every grade told us the story, and always ended (it must have been in the curriculum guide) by looking at us and saying, "So remember, children, if at first you don't succeed, TRY, TRY AGAIN."

Out of the teachers' hearing, some of my irreverent classmates used to laugh about Robert-the-Bruce and chant, "If at first you don't succeed, FRY, FRY A HEN." I never said fry, fry a hen. In fact, since I had been a failure twice before I was nine years old, I was interested in Robert-the-Bruce and thought about him a lot. But I decided I could never be like him. I was afraid of spiders, and if I were to be locked up in a dungeon with one, I wouldn't be inspired, I would cry. However, Robert-the-Bruce was surely a Success. He was dead.

On my own I worked out an ages-of-man theory. I decided that Successful people—those who had lived in a Golden Age, as it were—were all dead. The adults I knew who were alive were superior to me, but not Successes. Of course they could play golf and knew where buses went, but they didn't measure up to Robert-the-Bruce. I did not even know an adult who had been in a dungeon.

Pinky Higgins of the Chicago Cubs was an adult, but Grandma Annie was often cross with him, so he was not a Success. As for presidents, I had heard her saying some very plain things about President Roosevelt, so I knew that the affairs of our nation were in the hands of a Man of Brass.

Eventually, as is right and proper, I grew up. It was much better, even though I never became a Success. Instead, I married and had a son and became a librarian at Brown University in Rhode Island. Being a librarian has its points. You get to wear orthopedic shoes and a tiny frown as you snap the elastic band on your packet of catalog cards. Sometimes it is even exciting. I remember the day the assistant librarian, a rumpled, inept fellow, flushed the campus flasher from the stacks, chased him through the lobby and pinned him in the revolving door—but only because his own arm was caught in the door too. The campus cops came, took the flasher into custody and freed the assistant librarian. We talked about it during coffee break for weeks.

Paul was teaching at another university and directed the graduate biomedical engineering program there. He did important research, attended meetings and was a popular lecturer. He had been awarded tenure, that cachet of academic stability and ease. But he was not easy; in fact, he was often distinctly uneasy.

When that uneasiness became unbearable we quit our jobs and sold our house. Brian, whose absence had made the house seem so much bigger, was in boarding school, and we left Rhode Island and wandered around the country for nearly a year until we came to the Ozarks. I liked this farm the instant I saw it and Paul said he did too, so we bought it. We had to find something to do to make a living, and Paul said that since we didn't know anything about cows, we might as well become beekeepers. At the time that seemed perfectly reasonable.

As it turned out, I was luckier than he, for here I found what I wanted. Today, on our wedding anniversary, I think of him and hope that he has, too.

My chicken operation, I like to believe, is one of the few straightforward bits of farming that goes on at my place. But during the past weeks I have been trying to get the chickens organized to sleep inside the coop, and in doing so I've been forced to think like a chicken, which is not very straightforward at all.

The thirty chickens, give or take a few (especially take a few as coyote dinners), are a commercial strain of white Leghorns, egg-laying machines without the wit or attention to go broody. I sell their eggs back to the general store to cover the cost of their feed, have my own eggs for free and still have enough left over to keep Ermon's family across the hollow supplied in exchange for work on my truck.

Leghorns are skinny little chickens, useless as slaughter birds after they have stopped laying well. Some of mine die of old age; others are picked off by coyotes, foxes, hawks, raccoons, opossums and owls. Being a chicken on this hilltop is a perilous business. Each spring I start a dozen new pullets to compensate for the depredations on the flock, and the ones that make it past the black rat snakes reach egg-laying maturity in about twenty weeks. I buy sexed day-old pullets and brood them in the cabin, where they are safe and warm and I can watch them. This past spring it was apparent after a week that the chicken sexer had made a mistake.

Chicken sexing must be a highly skilled profession. I picture the sexers sitting there in the hatcheries, day after day, hour after hour, upending newly hatched chicks and sorting out the pullets from the cockerels. Considering how alike baby chicks are, they must be good at it, for they seldom make a mistake. But even Homer nods, and this past spring one of the baby chicks started

growing a tiny, bright-red comb, proving himself to be a cockerel.

Since the hens are temperamentally unsuited for raising their own chicks, I don't care if the eggs are fertile or not, and I don't really need to keep a rooster at all. But I like having one about, flapping his wings and crowing his exuberant maleness, and I fancy it to be a more wholesome state of affairs for the hens to have a rooster with them. I was glad to see the new cockerel this past spring because my old reigning rooster badly needed the lesson that only another young one could teach.

The old rooster is a magnificent New Hampshire red, twice as big as the Leghorn hens. His feathers shimmer in shades of red, brown and orange, and his tail, which is not quite so fine as he believes it to be, curves showily away from his body in deep coppery-green tones. His predecessor was a mild-mannered Leghorn who, one summer's day when drought had stressed all wildlife and their food sources, was carried off by a female coyote, long, lean, and hungry. She had chosen her time carefully, watching until the dogs had taken themselves for a walk to the river before she made her kill. I saw her, but it was too late. I chased her, but she looked disdainfully over her shoulder at me, the rooster dangling from her jaws, and broke into an easy lope, disappearing down into the creek hollow where I sometimes hear coyotes singing at night. My flock of hens was roosterless for a time, until I met a couple over on the next ridge of hills who had two roosters and one very tired hen, all of them New Hampshire reds. We arranged a trade: one of my laying hens for their extra rooster.

He is beautiful but mean, perhaps because he had been the number two rooster in the triangle, perhaps simply because the breed is aggressive. When I brought him home he stepped right out of the burlap bag and began setting things right among the Leghorns. He likes to keep all his hens in sight, pecking and chasing after them when they stray too far from the coop, scolding and fussing continually, full of self-importance. He attacks everything: odd-looking sticks, mice, cats, dogs, men, women and children, but particularly men, whose maleness he recognizes and whom, in some dim way, he may suspect of lusting after his hens. He might even take on a coyote, which, in addition to his beauty, is the reason I have put up with him for so long. But his personality is poor. He is a coward and a bully. If you face him, he backs down and discovers a delicious stone or speck that needs scratching, but turn your back and walk away and he comes running, fury in his wicked yellow eyes. His spurs are weapons, and whenever possible he jumps at a retreating figure, feet to the fore, and comes down stabbing with his spurs, his full ten pounds behind the attack. With a human he makes the connection on the back of the legs, and I bear my own set of scars there to show that, ungrateful brute, he will even attack the body that feeds him.

I knew when I saw the young cockerel among the chicks last spring that, proud and aggressive as he is, the aging New Hampshire red would be no match for the young Leghorn's inevitable challenge. By midsummer, when the newcomer first tried to crow, the old rooster knew it too. To me, the youngster sounded like a squeaky door hinge, but the old red rooster stopped in mid-bustle, transfixed, his head to one side. He watched the young rooster carefully after that and, pressing his temporary advantage of weight and experience, drove him out of the coop each night.

As a result, the Leghorn began to roost in the trees. Gradually most of the pullets joined him, and one by one so did the old laying hens. By summer's end, the red rooster, splendid in his possession of the safe chicken coop, had only six loyal hens. The ones who camped out always roosted too high for me to catch, but I did not worry about their safety, for they perched on flimsy branches where the raccoons could not follow, and the protective cover of leaves hid them from owls. But after the leaves fell this past autumn I was awakened several nights running by the frightened squawk of a chicken. By the time the dogs and I got outdoors there were only a few feathers left to show where the great horned owl, who calls his deep *hoo-hoohoo-hoo-hoo* from the pine tree, had made his kill.

Two weeks ago I set about getting the chickens back in the coop at night. They are impossible to catch, so I contrived to make them catch themselves, putting myself into a chicken frame of mind to do so. The chicken coop has two doors, a regular human-sized one and a small chicken-sized door. All the chickens, the ones who camp out as well as the coop birds, knew that I fed them just inside the big door and would gather nearby to watch me throw down the corn on the coop floor each morning. But when the outside birds went in, the inside ones came out, so I couldn't simply feed them inside and close the doors. I fashioned a tunnel of chicken wire and two-by-fours outside the small door, left it open at one end, and started feeding the chickens inside the makeshift pen. They were suspicious of the wire netting, but after a few days greed overcame caution and they went inside it to feed. On the first morning of my chicken trapping I left both doors closed and waited until later than usual to feed them. The outside chickens were gathered, hungry and anxious, and when I threw the corn into the tunnel, six of them, including the Leghorn rooster, rushed right in. I pulled shut a flap of chicken wire across the open end, then went inside the coop, fed those chickens, and opened the little chicken door. The coop chickens and the outsiders spent the day going back and forth between the coop and what was now a pen, the two roosters eying one another warily and keeping their distance. At day's end the outsiders reluctantly went to the coop and I closed the little door. The chickens outside the wire stayed hungry and roosted in the trees, uneasy without their rooster. The next morning most of them went into the tunnel for corn; since it was raining, all of them went inside the coop as soon as I had opened the little door, so I was able to close it again and keep them all together. I don't know what happened inside, but there was a lot of crowing and commotion, suggesting that the two roosters were beginning to sort out dominance. That evening only two hens remained to roost in the trees, and during the night the great horned owl carried off one of them. The last hen, deprived of company and corn for two days, was eager to get inside the chicken-wire tunnel the following morning, and at dusk I closed the entire flock into the coop. I fed and watered them inside, and left both doors closed for a week.

Chickens' tiny brains do not remember much, and when I let them out a few days ago they had lost all memory of roosting anywhere but in the coop, and tidily returned to it each evening. Rooster realignment had taken place. The mild and polite white Leghorn acts self-assured and is in charge; the red rooster, chastened and subdued, hasn't even the heart to spur me when I walk past him. I have heard the great horned owl each night in the pine tree close to my bedroom window. He will have to look elsewhere for dinner.

My farm lies north of town. After the first two miles, the black top gives way to a five-mile stretch of rocky road that shakes apart the pickups my neighbors and I drive. My mailbox is at the junction of this road and a mile-and-a-half gravel lane that meanders between it and the cabin, skirting the cliffs of the river that runs fast and clear below. Lichens, ferns and mosses grow there, and wind and rain have eroded caves and root holds for scrubby, twisted trees on the cliff faces. The thin soil at the top sustains a richer growth, and in the springtime the cliff top is abloom, first with serviceberry, then redbud and dogwood. In the summer, oaks shade the lane, grass grows in the middle of it, and black-eyed Susans grow beside it. In the winter, winds howl up out of the river gorge, driving snow across the lane in drifts so deep that sometimes I am marooned for a week or more.

I returned yesterday from a honey-selling trip and was grateful, as I always am, to turn at the mailbox and head down my lane. I drive a big three-quarter ton white truck on these trips, one fitted out to carry a 5,000-pound load, a truck new enough to be repaired if it should break down in Hackensack without hours of poking around in a salvage yard, the source of parts for "Press on Regardless."

The white truck is commodious and dependable, and I am fond of it. It is a part of my life. One night I dropped off to sleep after reading about the nature of the soul. I dreamed about my own soul, and found that it is a female white truck, buoyant, impatient, one that speeds along, almost too fast in an exhilarating way, skimming slightly above the road, not quite keeping to the pathway. I rather enjoy having a soul of that sort.

Like many of my neighbors, I am poor. I live on an income well below the poverty line—although it does not seem like poverty when the redbud and dogwood are in bloom together

—and when I travel I have to be careful about expenses. I eat in restaurants as little as possible, and I sleep in the truck: I pull into a truck stop, unroll my sleeping bag on the front seat and sleep there, as warm and comfortable as can be. In the morning I brush my teeth in the truck-stop restroom, and have my morning coffee in the restaurant. When I travel, people seldom notice or talk to me. I am unnoticeable in my ordinariness. If I were young and pretty, I might attract attention. But I am too old to be pretty, and rumpled besides, so I am invisible. This delights me, for I can sit in a booth at the truck stop, drink my coffee and watch without being watched.

One morning I was having coffee at 5:30 A.M., snugged up in a booth in a truck stop in New Mexico. The truckers were eating their breakfasts, straddling the stools at the horseshoe-shaped counter. A three-sided projection screen hung from the ceiling, showing slides that changed every minute or so. The truckers watched, absorbed, as the slides alternated between the animate and the inanimate. A supertruck, dazzling in the sunshine, every tailpipe and chrome strip gleaming, was followed by a D-cup woman, pouring out of her teeny dress, provocatively pumping gasoline into a truck. The next slide was a low shot of a truck grille; this was followed by a scene with a plump blonde in a cute cop outfit, showing rather more breast and crotch than one would think regulation, arresting a naughty trucker.

I watched the truckers as they watched the screen, chewing away on the leathery eggs-over-easy, their eyes glassy, as intent on chrome as on flesh. I finished my coffee and drove on unnoticed.

The trip I returned from yesterday was to Dallas, and as sales trips go, it was a good one. Its maze of freeways make it easy to get around, and I was grateful to the food buyers, who placed Texas-sized orders.

On the way to Dallas I stopped for lunch at an Oklahoma restaurant which had big windows facing the parking lot. Seeing the signs on my truck proclaiming my business and home town, the man at the cash register gave me a big grin when I walked inside and asked, "You the sweetest thing in Missouri?"

If there is one skill I have learned from living in the Ozarks, it is how to talk Good Old Boy, so I quickly replied, "Shore am," and took my seat at a table to order a bowl of soup. As I paid the tab, my new friend inquired about the honey business; when he found out that my truck was loaded with honey for sale in Dallas, he bought a case for the restaurant gift shop and asked to be put on my mailing list. "Now that's Joe Ben Ponder, you hear? Joe *Ben,*" he said in his soft southern Oklahoma drawl.

It seemed like an auspicious beginning for a sales trip, and I badly needed a good one. I had just returned from Boston and New York, where sales had been poor, although the trip was good in some ways. In Boston I stayed with Liddy and Brian, and one evening they took me to the Harvard chapel, where Gustav Leonhardt played a program of baroque music on the chapel organ. It was beautiful and I enjoyed it; I also enjoyed seeing other friends and relatives whom I love and see too seldom, but I did not make any money. In New York there are stores on every corner that sell French bread, marvelous cheeses, imported salmon, exquisite delicacies and honey, some of it made by my honeybees. But then there is another such store in the middle of the block. The customers are spread thin, and many places where I have sold honey for years have fired their managers and hired new ones, groping for a formula that will bring in the dollars once again. Macy's and Zabar's were having a war, and their buyers had no time for me. Sales elsewhere were poor, too, for it cut no French mustard with new managers that honey from my bees had been selling in the store for ten years. I drove up to Westchester and southern Connecticut to set up new accounts in the suburbs.

In my worn jeans and steel-toed work boots, one of which has a hole in it from the time I dripped battery acid on it, I wandered through those fashionable towns peddling honey, towns filled with women out buying things to drape on themselves, and things to put in their houses, and things to take care of the things hanging on themselves and the things in their houses.

Twenty or twenty-five years ago I lived on the edge of lives like these. In those days the women used to drive station wagons, and today they drive sleek little cars, but the look of strain on their faces is the same today as it was back then. I was glad to escape that life then and at the end of the sales day I was glad to escape in my white truck and head westward onto the Interstates with their green signs and truck stops, toward Missouri, toward my wild mountain top, toward home.

There has been an odd lot of siding and not-siding on the barn and its lean-tos: old weathered barn boards, tar paper, wooden shingles and, on the new storage loft, bare plywood. I needed to cover up this plywood before winter, and the handsomeness of the new loft suggested that now was the time to cover the barn and its sheds with something which would tie together the entire structure. The wooden shingles on the honey house lean-to had weathered beautifully, and I thought I might finish off the rest of the barn to match, but when I priced wooden shingles at the lumber yard they were impossibly expensive, so I made my own.

My raw materials came from the pallet mill to the east of town. Industrial pallets are made from four-foot oak blocks, trimmed to exact widths. It was the trim that made my shingles. The mill sells two thousand pounds of trim pieces for $2.00, my kind of price. I drove the Chevy over there, and a helpful mill worker loaded them into it with his forklift. When he dumped the bundle onto the pickup bed, the truck groaned a bit and squatted down, its springs splayed out flat. I drove home slowly and carefully, wondering how I would unload the bundle without having to cut the bands and take the trim pieces out bit by bit. But when I got home I put the truck to the task of unloading itself: I hooked a chain around the post that holds up the deck to the barn loft, fastened the other end to the steel bands around the pallet trim and drove out from under the load. The truck groaned in relief as the bed came back up to normal height.

On rainy days I finished putting up windows and closing openings in the new loft, and on sunny ones I shingled. Some of the trim pieces were either too thick or too thin, and these I put aside to use for kindling, but from most of them I could make three sixteen-inch shingles. I cut a stack on the table saw and then nailed them up, using a chalk line to keep the rows straight. The barn, to which Liddy and Brian's sculptured ridgebeam had given

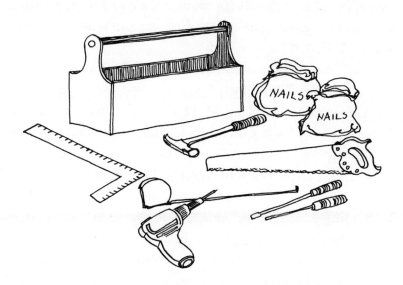

a head and tail, now, with its coat of tawny, rough, wooden shingles, looks like a shaggy beast squatting out there. I like it a lot.

I have made the bee equipment for my three-hundred hives myself, but real carpentry has always been beyond me. The skills that Liddy and Brian taught me and the practice I had on the barn went right to my head, and after the barn was finished I built a pump house with some help from a carpenter friend.

The pump house at the back of the cabin contained my well and pump. It was an ugly little building, all of its angles and proportions wrong. In addition, it has been rotting gently into the ground for five or six years. Last winter I draped a tarpaulin over the remains of the roof, but I knew that it would not last another one. It has stood there all this time, ill-seeming and an accusation each time I went out the door.

The location of the well meant that the pump house would have to be in an obtrusive spot. I had asked Liddy and Brian what I could do to make a new one look better. I wondered about an asymmetrical roof, but Brian cautioned me against it, pointing out that the buildings around the farm were all symmetrical and balanced.

"Build something asymmetrical in this setting and it will look like Quasimodo," he said.

In my mind I worked out what I wanted, and with a carpenter friend's help we built a neat little building. But when we came to the peaked roof, we found that we had to add a second, higher, removable, overlapping roof to one side to allow space for the pump to be pulled out in case repairs are needed. I have indeed built Quasimodo, but as with Quasimodo, it is likable. It is smaller, better-proportioned than the old pump house, sided with weathered barn siding, trimmed in International Harvester red which highlights the faded red of the barn boards, and roofed and hump-roofed in ripply tin. It is a cheerful and satisfying little building to have outside my back door, a good companion to the great shaggy barn beyond.

Buoyed by the prospect of being able to build something that serves its purpose and looks the way I want it to look, I repaired the chicken coop and put it to rights, added several features to the honey house and main part of the barn that I have always wanted, and then, as the weather became colder, moved inside the cabin to finish my office, a place of studs, very little wallboard and half a ceiling. Finishing that was so exhilarating that I turned to the bedroom. Like the office, it has stood unfinished since Paul left, and in the past two weeks I have ripped it apart and rebuilt it in a rather massive way. Yesterday I finished and carried all the tools back out to the barn: saw horses, circular saw, variable speed drill and driver, carpenter's level and square, tape measure and tin cans full of screws and nails.

Then I cleaned up the bedroom and rearranged the furniture. When I went to bed last night to read, as I usually do, I realized after awhile that I was not reading the book propped up on my knees at all. Instead, peering over the top of my glasses, I had been admiring the new room around me and had been lost in reverie, smiling foolishly and thinking with gratitude of what Liddy and Brian had given me with their carpentry lessons.

WINTER

I am rich in birds today, especially in bluebirds and cedar waxwings. The bluebirds—eastern bluebirds, to be precise—nest here less and less during the summer but congregate in flocks in winter. Brilliant blue, with rust-colored breasts and white bellies, they have been here for over a month eating sumac berries and the fruit of the pepperidge trees that grow all around the cabin, chittering contentedly to one another as they feed. The waxwings, however, are erratic visitors, and never before have I had them in such numbers or in such steady attendance. I have been seeing and hearing them for several weeks now.

Today, after three days of rain and sleet which made it hard for them to feed, the sun rose in a clear sky, and not long afterward I could hear the *tsee-tsee* of the cedar waxwings in the trees. I had spent the past days of sleety weather inside the cabin labeling honey jars, a necessary but boring job, and I too was glad to see the sunshine. I put on a jacket, hung my binoculars around my neck and went out to watch the waxwings. There must have been hundreds of them visible in the bare treetops. I could hear them everywhere. The bluebirds were perched on the power lines

and separately in their own trees, although here and there I could see one in among the cedar waxwings. Goldfinches in dull winter plumage were flying between trees in swooping flight, singing as they do in summertime on each dip. Cardinals, chickadees and nuthatches were jostling one another at the feeder. On the ground, which was still covered with frost in the shadows of the barn and outbuildings, juncos—small, trim, neat gray-and-white birds—were scratching for weed and grass seeds so earnestly that they did not even fly up as I walked among them.

I went out into the field. Bluebirds were singing on all sides, calling back and forth quizzically: *chur-wee? chur-wee?* I could also hear cedar waxwings from over by the barn, and through my binoculars I could see that they were settling down to feed in the little persimmon grove. The persimmon trees flower in

June, when the bees work the blossoms for nectar in such buzzing numbers that when I am walking in the woods I can hear a persimmon grove long before I see it. The fruit the bees help to set is smaller than the Oriental persimmons found in specialty markets, but it is tasty and high in sugar content after it is ripe. It ripens late, not until after frost; before that it is bitter and acrid. I once bit into one too soon. My mouth puckered immediately, and the astringent effect lingered on into the next day. But ripe persimmons are sweet and flavorful, and I had intended to shake down the fruit of those trees in back of the barn and make a pie from them for a couple who are friends. The wife is an excellent cook but she told me once that pies are beyond her, although she likes to eat them. I am no cook at all, but I rather fancy making pies when I have someone to make them for because they look so pretty when they are finished. I was invited to dinner with my friends and was going to take them a persimmon hickory-nut pie as a surprise. But the cedar waxwings needed the persimmons more than my friends needed a pie, so I am glad I never got around to shaking the fruit from the trees.

I went up into the barn loft for a closer look. The new storage loft put me right at the level of the tops of the young trees, and the windows allowed me a close look without disturbing the birds. They were feeding so greedily, however, that I doubt if I could have disturbed them anyway.

Cedar waxwings are sleek, elegant birds, crested and brownish-backed. Their bellies are creamy white. A bright band of yellow tips their tails, and their wings are banded in red. A black band runs from beak top around their eyes to the back of their head, giving the party in the persimmon trees the air of a masked ball. One of my bird books calls them, fancifully, "birds of mystery." This is because they are wanderers. Out of breeding season they are social birds who congregate in such big groups that they quickly deplete the food in any one spot, so they are always on the move. They appear one day in numbers, eat everything— insects when they can find them, fruit of any kind, dogwood berries and, like the bluebirds, sumac and pepperidge—and then fly on, not to be seen again. Their presence here for nearly a month seems highly unusual, but this was a good year in these parts for the kinds of things that cedar waxwings like to feed on and even now there is still plenty for them to eat.

Thomas Nuttall, the nineteenth-century eccentric whose name was given to more species of plants and animals than any other naturalist, claimed cedar waxwings were excessively polite to one another. He said that he had often seen them passing a worm back and forth down a whole row of beaks before it was finally eaten. I would enjoy seeing this, but I never have. Today the birds I was watching from the loft were ravenously hungry after the days of bad weather, and though they did not take persimmons from one another so that in a sense their behavior might be called "polite" they did not pass food around, either. Instead, each bird chose a single persimmon, tore at the skin to get to the sweet soft pulp inside, gobbled it down in big lumps, and then hopped to another free persimmon to gorge again. They were feeding so intently that they stopped calling to one another. I watched them for a long time, until, their hunger stilled, they flew off to a different tree, where I could see them through the binoculars preening and sunning and hear them socializing sibilantly, *tsee-tsee-tsee*, in a satisfied way.

This is the recipe for the pie I would have made:

1 graham-cracker crumb crust
1/2 cup brown sugar
1 envelope unflavored gelatin

1/2 teaspoon salt
3 eggs, separated
2/3 cup milk
1 cup strained wild persimmon pulp
1/4 cup sugar
1/4 cup hickory nuts, chopped
Heavy cream for whipping

In a saucepan combine the brown sugar, gelatin and salt. Beat the egg yolks lightly and add to the milk. Stir into the brown sugar mixture and cook, stirring constantly, until mixture is thoroughly heated and sugar is melted. Avoid boiling. Remove from heat and stir in persimmon pulp. Chill for approximately one hour or until soft mounds form when mixture is dropped from a spoon.

Beat the egg whites until soft peaks appear. Add the sugar gradually, beating until stiff. Stir the hickory nuts into the chilled persimmon mixture and fold into the beaten egg whites. Turn into crumb crust and chill until firm.

Serve with sweetened whipped cream.

The best place to store firewood is in an airy shed, protected from rain and snow. I wish I had such a shed but I don't, so I stack my firewood out in the open down below the barn. Each morning in the late fall and winter, I split enough wood for the second day, carry it to the cabin and pile it near the wood stove where it can dry.

As the wood warms, the wood cockroaches that sometimes live under the bark are roused and creep out. They are smaller than the American cockroaches that, even in the tidiest of houses, invade kitchens at night looking for scraps of food, but they are bigger than the familiar German cockroaches that live in most kitchens and bathrooms. German cockroaches are called Croton bugs in the northeast because they were first noticed in New York in the mid-1800s, when the Croton aqueduct was built. It is a safe bet that the Croton bugs noticed human beings well before then, for roaches sized us up long ago as the providers of good habitat. Their relationship has been so close that there probably is not a human alive so entomologically ignorant that he cannot identify a cockroach when he sees one.

My wood roaches, which come in several species, are dark

brown and shiny, and are seldom a problem for human beings.
They usually live outside houses and eat wood, a diet made
possible by the fact that their gut is inhabited by the same
intestinal protozoans which allow termites to digest wood. This
is one of the reasons termites and cockroaches are believed to have
a common ancestor somewhere in the distant past.

Cockroaches are possibly the most successful complex life
form this planet has ever seen. Geological dates are inexact, but
will serve for comparison. Humans have been around for perhaps
two million years. Honeybees, along with flowering plants,
evolved sometime during the Cretaceous period, some 100 million years ago. But before that, back in Upper Carboniferous
times, 250 million years ago, fossil records show that roaches—
at least 800 different kinds of them—were scuttling about. On
the average, they were somewhat bigger than any of the 1200

species of roaches that exist now, but none of the fossil forms are
any bigger than contemporary tropical ones. The differences
between those fossil cockroaches and the ones of today are slight,
and mainly involve the position of the wing veins.

The world was quite a different place in those days: warmer, moister and filled with animals and plants that would seem strange to us. The ancient seas washed on different beaches. The waters had started to recede from the central part of this continent, and in the swamps and marshes along the shores grew giant horsetails, ferns, mosses and primitive conifers. There were no flowers. Huge dragonflies filled the air. There were no birds. Primitive amphibians and reptiles walked the land, and cockroaches scuttled about, feeding on plant and animal materials and debris just as they do today.

Climate, land masses, plants and animals changed through the ages, but not the cockroaches. They had found a form even then that worked exceedingly well, a form perfect for survival on this planet, and natural selection doesn't argue with success.

They were and are neither too big nor too small. Their low-slung, flattened bodies allow them to hide in the merest sliver of protective space. They could and can live in a wide variety of habitats and temperatures, and eat almost anything. They are tough, spunky, wary and alert all of their lives. Young roaches hatch from leathery egg cases ready to run. Like their cousins the grasshoppers, katydids and crickets, they do not undergo a complete metamorphosis. They do not spend periods of their lives as vulnerable, wormlike larvae or quiet pupae rearranging their body tissue and structures. Instead, they undergo a gradual metamorphosis. Young hatchling roaches look very much like adults, except that they are tiny and lack wings. They shed their outer skins several times before they reach reproductive adulthood, but at every stage they are quick and agile. Adults can fly, but more typically run.

They have never become specialized; they have always been highly flexible, able to adapt to any changes that the world has in store. I was not surprised when I read that a researcher had discovered cockroaches could learn things even after their heads had been cut off, for back when I first started keeping bees if I opened a beehive and found roaches inside, believing myself to be the bees' ally, I would cut the cockroaches in two with my hive tool. Invariably, the nether end would scurry off, apparently able to function quite nicely without the head it had left behind.

There are always American cockroaches in beehives. Like human houses, beehives are warm, snug places, well-stocked with food, and roaches live there if they can get away with it. When a colony of bees is active, healthy and strong, the bees will not tolerate roaches any more willingly than does a fussy housekeeper. I have often watched honeybees chasing cockroaches out of their hives, and have also seen them carrying out roach egg cases and dropping them some distance away, recognizing that they are objects inappropriate to a well-regulated bee colony. There is constant strife between the two species. The bees are vigilant and aggressive, but the roaches are always there, and at the least drop in hive strength or morale, they take over. They are opportunists.

In the past few years, I have left off killing cockroaches when I open a beehive. I now know that a good colony of bees can take care of them on their own better than I can. And if a colony is not a good one, I had better find out what is wrong with it rather than kill its roaches.

In truth, I don't mind the wood cockroaches that come in on my firewood, either. Their digestive system and mine differ enough so that we don't share the same ecological niche; they do me no harm, we are not competing, so I can take a long view of them. There is no need to harry them as a bee would, or to squash them as a housewife would. Instead, I stoop down beside them and take a closer look, examining them carefully. After all, having in my cabin a harmless visitor whose structure evolution has barely touched since Upper Carboniferous days strikes me, a representative of an upstart and tentative experiment in living form, as a highly instructive event. Two hundred and fifty million years, after all, is a very long view indeed.

Winter 138

A group of people concerned about a proposal to dam the river came over to my place last evening to talk. The first to arrive was my nearest neighbor. He burst excitedly into the cabin, asking me to bring a flashlight and come back to his pickup; he had something to show me. I followed him to his truck, where he took the flashlight and switched it on to reveal a newly killed bobcat stretched out in the bed of his truck. The bobcat was a small one, probably a female. Her broad face was set off by longer hair behind her jaws, and her pointed ears ended in short tufts of fur. Her tawny winter coat, heavy and full, was spotted with black, and her short stubby tail had black bars. Her body was beginning to stiffen in death, and I noticed a small trickle of blood from her nostrils.

"They pay thirty-five dollars a pelt now over at the county seat," my neighbor explained. "That's groceries for next week," he said proudly. None of us back here on the river has much money, and an opportunity to make next week's grocery money was fortunate for him, I knew. "And I guess you'll thank me because that's surely the varmint that's been getting your chickens," he added, for I had said nothing yet.

But I wasn't grateful. I was shocked and sad in a way that my neighbor would not have understood.

I had not heard a shot and didn't see the gun that he usually carries in the rack in his pickup, so I asked him how he had killed her.

"It was just standing there in the headlights when I turned the corner before your place," he said, "so I rammed it with the pickup bumper and knocked it out, and then I got out and finished it off with the tire iron."

His method of killing sounds more savage than it probably was. Animals in slaughterhouses are stunned before they are killed. Once stunned, the important thing was to kill the bobcat quickly, and I am sure my neighbor did so, for he is a practiced hunter.

Others began to arrive at the meeting and took note of the kill. One of them, a trapper, said that the going price of $35 a pelt was a good one. Not many years ago, the pelt price was under $2. Demand for the fur, formerly scorned for its poor quality, was created by a ban on imported cat fur and a continuing market for fur coats and trim.

My neighbor and the trapper are both third-generation Ozarkers. They could have gone away from here after high school, as did many of their classmates, and made easy money in the cities, but they stayed because they love the land. This brings us together in our opposition to damming the river to create a recreational lake, but our sensibilities are different, the product of different personalities and backgrounds. They come from families who have lived off the land from necessity; they have a deep practical knowledge of it and better skills than I have for living here with very little money. The land, the woods and the rivers, and all that are in and on them are resources to be used for those who have the knowledge and skills. They can cut and sell timber, clear the land for pasture, sell the gravel from the river. Ozarkers pick up wild black walnuts and sell them to the food-processing companies that bring hulling machines to town in October. There are fur buyers, too, so they trap animals and sell the pelts. These Ozarkers do not question the happy fact that they are at the top of the food chain, but kill to eat what swims in the river and walks in the woods, and accept as a matter of course that it takes life to maintain life. In this they are more responsible than I am; I buy my meat in neat sanitized packages from the grocery store.

Troubled by this a few years back, I raised a dozen chickens as meat birds, then killed and dressed the lot, but found that killing chicken Number Twelve was no easier than killing chicken Number One. I didn't like taking responsibility for killing my own meat, and went back to buying it at the grocery store. I concluded sourly that righteousness and consistency are not my strong points, since it bothered me not at all to pull a carrot from the garden, an act quite as life-ending as shooting a deer.

I love this land, too, and I was grateful that we could all come together to stop it from being destroyed by an artificial lake. But my aesthetic is a different one, and comes from having lived in places where beauty, plants and animals are gone, so I place a different value on what remains than do my Ozark friends and neighbors. Others at the meeting last night had lived at one time in cities, and shared my prejudices. In our arrogance, we sometimes tell one another that we are taking a longer view. But in the very long run I'm not so sure, and as in most lofty matters, like my failed meat project, I suspect that all our opinions are simply an expression of a personal sense of what is fitting and proper.

Certainly my reaction to seeing the dead bobcat was personal. I knew that bobcat, and she probably knew me somewhat better, for she would have been a more careful observer than I.

Four or five years ago, a man from town told me he had seen a mountain lion on Pigeon Hawk Bluff, the cliffs above the river just to the west of my place. There is a rocky outcropping there, and he had left his car on the road and walked out to it to look at the river two hundred and fifty feet below. He could see a dead turkey lying on a rock shelf, and climbed down to take a closer look. As he reached out to pick up the bird, he was attacked by a mountain lion who came out of a small cave he had not been able to see from above. He showed me the marks along his forearm—scars, he claimed, where the mountain lion had raked him before he could scramble away. There were marks on his arm, to be sure, but I don't know that a mountain lion or any other animal put them there. I suspect that the story was an Ozark stretcher, for the teller, who logs in many hours with the good old boys at the café in town, is a heavy and slow-moving man; it is hard to imagine him climbing nimbly up or down a steep rock face. Nor would I trust his identification of a mountain lion, an animal more talked of at the café than ever seen in this country.

Mountain lions are large, slender, brownish cats with long tails and small rounded ears. This area used to be part of their range, but as men moved in to cut timber and hunt deer, the cats' chief prey, their habitat was destroyed and they retreated to the west and south. Today they are seen regularly in Arkansas, but now and again there are reports of mountain lions in this part of the Ozarks. With the deer population growing, as it has in recent years under the Department of Conservation's supervision, wild-life biologists say that mountain lions will return to rocky and remote places to feed on them.

After the man told me his story, I watched around Pigeon Hawk Bluff on the outside chance that he might really have seen a mountain lion but in the years since I have never spotted one.

I did, however, see a bobcat one evening, near the rock outcropping. This part of the Ozarks is still considered a normal part of bobcat range, but they are threatened by the same destruction of habitat that pushed the mountain lion back to wilder places, and they are uncommon.

Bobcats also kill and feed on deer, but for the most part they eat smaller animals: mice, squirrels, opossums, turkey, quail and perhaps some of my chickens. They are night hunters, and seek out caves or other suitable shelters during the day. In breeding season, the females often chose a rocky cliff cave as a den. I never saw the bobcat's den, but it may have been the cave below the lookout point on the road, although that seems a trifle public for a bobcat's taste. The cliff is studded with other caves of many sizes, and most are inaccessible to all but the most sure-footed. I saw the bobcat several times after that, walking silently along the cliff's edge at dusk. Sometimes in the evening I heard the piercing scream of a bobcat from that direction, and once, coming home late at night, I caught her in the road in the pickup's headlight beam. She stood there, blinded, until I switched off the headlights. Then she padded away into the shadows.

That stretch of land along the river, with its thickets, rocky cliffs and no human houses, would make as good a home ground as any for a bobcat. Females are more particular about their five miles or so of territory than are males, who sometimes intrude upon one another's bigger personal ranges, but bobcats all mark their territories and have little contact with other adults during their ten years or so of life.

I don't know for sure that the bobcat I have seen and heard over the past several years was always the same one, but it probably was, and last night probably I saw her dead in the back of my neighbor's pickup truck.

In a few days it will be Christmas. I have been snowed in for several days now, which is of no account. Toward the end of December I always go into town and stock up on extra feed for the chickens, dogs, cat, wild birds and supplies for myself, for there is usually a week or more during January when the lane to the mailbox is impassable to a truck because of ice or snow. The weather came a bit early this year, but I have already stocked up and do not need to get out.

The lane is a county road, and after he has cleared the school bus routes, the man who drives the township grader may clear my way. But the township is a poor one and the grader is old and undependable and held together with baling wire and ingenuity so he may not get here before the snow melts.

A few years ago, after another heavy snow, the man cleared the lane. It was bitterly cold that day, so when I heard him coming I went down to the end of the driveway and invited him in for coffee. He accepted, drove the grader up to the cabin, and turned it off. Instantly a look of distress swept across his face. He had forgotten that his battery was not holding a charge, and now that the engine was off he would never get it started. But I have a battery charger and plenty of extension cord, so while the coffee percolated he hitched the grader up to the charger. We sat by the wood stove talking of roads and weather, and by the time we had emptied the coffee pot, the battery was charged enough to start the engine. Country living requires cooperation.

Fifteen years ago, when I was working as a librarian at Brown University, I had a forty-five-minute drive to work each day. Before that I worked at a state college in New Jersey and had an hour-long commute. I grew to loathe winter, dreading the drive on slippery, congested highways. Winter was an enemy I had to fight. But no longer. I plan my sales trips so that I am off the road in bad weather. Instead, I repair equipment, label honey jars, prepare for the spring bee season, and generally work around here in the barn or cabin. Winter is not an enemy. It is a time of less going about and brings quiet and peace.

The mailman has not been able to get through on the back roads for several days, but he telephoned to tell me that he is going to try to make it today, so I shall walk down to the mailbox later on to see what he has left. It is a walk that I enjoy at any time of the year. Tazzie and Andy like it too, and in more benign weather they rush on ahead joyfully. When the drifts are hip high, as they are today where the lane skirts the river's cliffs at Pigeon Hawk Bluff, I try to urge them to break a path, but they look at me wisely and pretend that they are too loyal and obedient to do anything but walk at my heels. Pantywaist dogs, I scold. This makes them wag their tails happily. Are they my dogs or am I their human?

Brian telephoned last night from Boston and asked what I was going to do for Christmas. Not much, I admitted; what was he going to do? Well, he certainly hoped to have a Christmas tree, and so should I. I protested against cutting a tree, bringing it into the cabin and covering it with little shiny things. No, of course that would be inappropriate, he said, but certainly with all those trees around, one of them might be a Christmas tree, mightn't it?

He was right, as he usually is. So I declared the pine tree outside the three big windows to be my Christmas tree and hung suet on it as a gift to the blue jays, nuthatches and red-bellied woodpeckers. The woodpeckers have already found it.

The feeder with birdseed in it goes all the way across the windows, and the usual winter birds are feeding at it: juncos, cardinals, titmice, tree sparrows and finches, both purple and gold. The red-bellied woodpeckers like it too, and when they

come so close I can see the faint pencil-thin streak of red on their bellies that gives them their name. This morning I counted eight eastern bluebirds up on the power line. They do not come to the feeder because they are not seed eaters, but they congregate where there are other birds and feed on the sumac and dogwood berries they find at the edge of the field now that they have stripped the pepperidge trees.

Inside, Tazzie, Andy and Black Edith are luxuriating in the warmth of the wood stove. The public radio station is outdoing itself in programming during Christmas week. Last night I heard the Bach B Minor Mass, and today they have promised a program of Renaissance Christmas music. It feels snug and cheerful and peaceful here.

The secretary of the park superintendent just telephoned to cancel an appointment I was supposed to keep with him this afternoon concerning the dam that my neighbors and I are worried about. A group of local people, egged on by political and commercial interests, are requesting that a dam be built on the river just below my place on the federal land. Several thousand people have signed a petition offering the river and themselves to the Corps of Engineers as a sacrifice: an environmental, economic and personal disaster. It is also an absurdity, but politics has a way of turning the absurd into reality, so I have to pay attention. I do not think that dams get built by two thousand people asking their congressman for one; on the other hand, I do not know how they *do* get built, so I have been making it my business to find out. My appointment with the superintendent was part of this business.

However, neither he nor I can keep the appointment today. Winter on the banks of the river has canceled it, and mocks the making of plans. It is not a good time of year for political activism anyway; it is a time for privacy and indwelling. I am going to go out now and split firewood enough for a couple of days, and bring it in to dry beside the wood stove. Then I shall break up bits of a dead branch from the oak that grows by the driveway for kindling tomorrow morning's fire. In wintertime on the river I think this is as far ahead as it is wise to plan.

Yesterday was one of those clear bright days that we often have here in winter in the Ozarks. The temperature rose enough to let the icicles on the chicken coop begin to melt. The blue sky and bright sunshine drew me out of the cabin, where I had been dutifully labeling honey jars for several days. I proposed a walk to the dogs, and we headed out across the field to look at the beehives back by the woodlot.

The dogs ran on ahead, stopping briefly to snuffle excitedly where a pair of wild turkeys have been scratching and feeding during the past week just beyond the barn. All three of us had watched them from the windows, some of us more noisily than others.

The snowy roadway across the field was covered with footprints: deer, turkey, rabbits, woodrats and mice had all been forced out in recent days to look for food, and their tracks crisscrossed the path. For the past two days I have been seeing an uncommon number of hawks, both red-tailed and rough-legged, hunting in the field for rats and mice. At one spot I found a ragged depression in the snow stained with blood. Bits of down suggested that an owl had taken a victim there during the night.

When I got back to the beehives, all appeared quiet. It was still far too cold for bees to fly. I hoped that they had enough honey to feed on for the winter. If they did they were alive, clustered inside their hives, metabolizing fiercely, fueled by honey, to keep the temperature in the cluster at seventy degrees. Now that the days are beginning to lengthen, they will raise that temperature inside the brood area to ninety degrees as the queen starts to lay eggs. The cycle begins anew.

To open a hive and check them would be cruel in these temperatures, for all their generated heat would escape and the seals that they have so carefully made to keep it in would be broken. Still, by walking in front of the hives, I could tell about the health of the colony within. The bees in strong and able hives had been tending to sanitation, flying out to defecate, spattering the snow with their yellow droppings, and carrying out the corpses of some few sister bees who had died of age or cold.

Walking along looking for these signs, I discovered a young opossum crouched between two beehives. Opossums, and skunks too, can be a nuisance to bees in the winter. The bees are too

lethargic to defend themselves, and the opossums reach their forepaws inside the hive, stir up the cluster, capture bees one at a time and suck them dry of their honey and soft body parts. I once came upon a full and contented opossum sitting in front of a beehive ringed with husks of bee bodies, looking like a human glutton who had just overeaten at a clambake.

The opossum I found yesterday had not begun to feed. He was young and looked inexperienced. He was small, no bigger than a kitten, and frightened of me. He opened his mouth wide and tried to look fierce as an adult would do, but his teeth were still small.

There is only one species of opossum in North America, *Didelphis virginiana,* and this one, like the rest of his kind, had an untidy grayish coat, a naked, prehensile tail, and deceptively soft-looking pink feet. His blackish ears were edged in pink too. Opossums, this hemisphere's only marsupials, are among the most primitive of living mammals, and this one, with his head back and his jaws open, had a prehistoric look. Opossums eat fruit, nuts, insects, carrion and an occasional chicken when they can get one. In the Ozarks they are considered vermin, and are sometimes hunted for sport and trapped for their fur.

Some weeks ago I found an adult opossum in the chicken coop. I had returned home late after spending an evening with friends. The dogs greeted me as joyfully as though I had been gone for weeks, and leaped about as I walked over to gather the eggs and close the chicken-coop door. When I switched on the light in the coop we could see the opossum in a corner under the droppings board. His mouth was wide open, displaying formidable teeth, and he was ready to fight. The dogs often take themselves off on their own hunting expeditions and they must know how fierce an opossum can be, for they quieted down immediately and discovered that they had important business over by the cabin door, where they waited circumspectly for me. I threw a piece of kindling at the opossum to drive him from the coop.

With the memory of this encounter in mind I didn't expect much action when the dogs came looking for me as I stood in front of the beehives. The opossum saw the dogs before they saw him. He tensed his body and opened his mouth even wider, drooling with the effort. A tiny, choked growl warned the dogs to come no closer, but also attracted their attention. Tazzie must have decided that he was a cat and of no account, for she walked behind the opossum and upended him with her nose, sniffing his rear as she does with the cat, a behavior that appears to be both affectionate and curious. The opossum had probably never met with such an indignity and was confused.

Andy is an old, experienced hunter. He knew that this was no cat and rapidly decided that this particular creature was not as fierce as others of its kind with which he had tangled. He took advantage of the momentary confusion, and, surprising me with his quickness, rushed the opossum and grabbed him by the back. I called him off, but once he has caught an animal he is an efficient killer, and it appeared that he had already snapped the opossum's backbone. Both dogs sat beside me obediently, and we all looked at the opossum, quiet and still, lying in the snow.

I called the dogs to heel and we walked back to the cabin. I was filled with remorse for having stood there and allowed the dogs to bait a young nocturnal animal so desperately hungry that he had been driven out in the daylight in search of food.

After shutting the dogs in the cabin, I walked back to the beehives. When I got there the opossum was gone. I could see the tracks in the snow where he had run from the hives to the safety of the woods and thickets. He had played possum, gone limp, feigned death, practiced the ultimate in passive resistance.

I returned to the cabin sobered, thinking about what had happened as I walked along the roadway where most of the animal tracks had been obliterated by my own and those of the dogs.

I do not like opossums to hustle their dinners in my chicken coop or beehives, but that is the way creatures get on, and I would not kill one for doing so, Ozark custom to the contrary. I was relieved that the young opossum still lived. His life had touched mine on a bright and sunny day, and I was grateful for the encounter. I would have been sad and guilty had he lost it because of our contact. I hope he found something to eat after all, and fed well.

The coyotes are singing almost every night now down in the hollow, the creek bed that runs below my southern boundary. Their song is a chorus of yips, yelps and barks that increases in intensity and number as more and more join in, until it climaxes in a series of howled wails. I hear them year around occasionally down there or over by the river, but it is their breeding season now and they are singing more often.

When I first moved here, their song was rarer. The pair of male Irish setters Paul and I had then would cock their heads and listen when they heard coyotes, and then look at us as if for an explanation of this sound so near and yet not quite of their kind.

The dogs I have now, Tazzie and Andy Beagle, are accustomed to the coyotes, however, for here as elsewhere coyotes have increased in number. They are one of the few wild animals who have spread their range, even as humans have tried to eliminate them. They are shrewd and wily and can eat almost anything. Here in Missouri their diet is chiefly rabbits, rats, mice and small birds, but they will eat insects, plants, fruits and seeds when they have to. They bring back their prey to their den and bury what they don't eat to dig up later when they need it. They are fond of carrion wherever they find it; they will polish off the kill of another animal and clean up after inexpert human hunters who only maim. Though studies of the stomach contents of killed coyotes show that only ten to twenty percent of what they eat are animals man would prefer they didn't—calves, sheep or chickens—ranchers and farmers consider them enemies. Bounties have been placed on their heads, traps and poison put out for them. Still they spread. Back here on the river I am the only human who likes to hear their song.

A neighbor told me that in some circles a coyote-dog cross is held to be a superior hunter. He knew a man who had trapped a female coyote in season and held her, snapping and furious, while his prize hound covered her. He kept the coyote caged until she whelped, but she killed her pups as soon as they were born. I am not surprised after such a mating and imprisonment. I am also not sure that the story is true, but it could be. *Canis familiaris,* our pet dogs, and *Canis latrans,* the dog who barks, the coyote, are close enough biologically to interbreed. They sometimes do, and produce fertile offspring.

I suspect that Andy Beagle knows more about this than he should. In his younger days he would go quite wild for a week or so in late January and spend his nights courting, limping back home in the morning, mauled and bloody. He would stay some hours by the wood stove whimpering for my sympathy and licking his wounds. By afternoon he would be restless again, looking out the windows and whining. He would sit in the middle of the living-room floor, point his nose to the ceiling and howl his randiness, and by evening was ready to hobble out of the cabin and away, only to return again the next morning, his old wounds reopened and new ones bleeding.

I always asked around to see if any of my neighbors had a bitch in heat, but none of them ever did, and several of them suggested that he was courting coyotes. It is possible; if so, he carries the scars of his chutzpah: one long floppy ear in three parts, rather like a split-leafed philodendron, the badge of a night when he returned covered from nose to tail with blood that was streaming from the ripped ear. It was the last time he ever went wild in January. He is older now; he spends the month as close to the wood stove as he can get.

Perhaps Andy learned something in his final wooing. I did see one encounter of his with a pair of coyotes in which he showed himself discreet if not valorous—prudent behavior, which indicated to me that he was well acquainted with coyotes and their capabilities. Coyotes are not as sociable as their near cousins, wolves, but they do often hunt in pairs and are sometimes seen in daylight in the summer after their pups have become active. Our meeting with them was on a summer day when I was stacking logs on the woodpile out beyond the barn. Andy had joined me and had discovered the trail of some small animal that had taken refuge there. Nose to the ground, intent on whatever the scent was telling him, he followed the trail out across the field. I straightened up from behind the stacked wood and looked in his direction as two coyotes emerged silently from the woods at the edge of the field above the hollow. Bellies to the ground, they were stalking him but he was following a scent and was unaware of them until they began to circle. Then he sensed their presence, and looked up and yelped in fear. They circled closer. He turned his head to show that he was not aggressive, and would have backed off if they had given him room. The coyotes tightened their circle. Andy lay down, rolled over and exposed his belly to show his submission. The coyotes surely understood these gestures, for like dogs, they use them among themselves to sort out dominance, but social niceties were not on their minds; they were hunting. They closed in on him, snarling, showing their teeth, snapping. When I yelled, they looked up, momentarily startled, and in that moment Andy jumped up and streaked back to the cabin with his tail between his legs. I chased the coyotes, and although they did lope away toward the hollow, they stopped often to look insolently over their shoulders at me. I had won this round, their behavior said, but they were not frightened of me.

Now that Tazzie has joined the household, she and Andy always go out on their rambles together, and the two of them may be a match for coyotes. I do not know, although I have seen them chase a lone coyote. I am sure, however, that they know more about them than I do. When they hear coyotes singing at night they do not bark like they do when they hear dogs, but they do not act puzzled the way the setters did, either. They listen attentively, file away whatever information they get from the song and go back to sleep: *Coyotes. Nothing special.*

But for me, perhaps because I know them less well, coyotes are still special. I like lying in bed as I did last night, with moonlight streaming in the window, listening to their song. Coyote. That name comes from the Aztec's word for them, *coyotl,* and should, to my way of thinking, always be pronounced ki-o-tee, not ki-oat, the way it sometimes is. Coy-o-tee, that word handed down from long ago, a clumsy approximation of their song, the song of wild things in the moonlight.

Every winter I promise myself that I will label an entire year's supply of honey jars, and every winter I don't; the job is a boring one, and I am clever about finding other work more important.

I sell most of my honey in one-pound jars. There are twenty-four of these to a case, and each one needs a label. I set up my label-pasting machine on a table near the wood stove. The machine is so noisy that I can't listen to the radio, and although I don't mind because my public radio station has recently come down with a bad case of Dvořák, the labeling machine doesn't have much interesting to say either—just *thwip-thwip-thwip* as the labels run through its rollers. Sitting there by the warm fire pasting label after label neatly and precisely makes me sleepy and dull.

But I did get enough jars labeled for the February sales trips, and now I am processing honey to fill them.

When I extract the honey at summer's end I store it all in five-gallon, sixty-pound buckets and stack them four high in the barn. This honey has never been heated, and so it has crystallized. I wish I could sell it that way because the flavor is at its best, and the thick crystallized honey spreads nicely on warm toast without running down the elbow the way liquefied honey does. But the buyers at stores tell me that their customers think something is wrong with honey when it has crystallized, so I must melt it down and heat it before I can sell it.

I put ten of those sixty-pound buckets into a hot box, which melts the honey so that it can flow out of the buckets and into a pump, which pumps it up into an overhead controlled heating unit. This heats it enough to break down the honey crystals but does not hurt the flavor. The honey flows from the heater down into a storage tank, where I let it stand to allow the air bubbles to rise out before I bottle it. Processing each batch of six hundred pounds of honey takes one day.

The bees made the honey from flower nectar. Bees forage as far as two miles from their hives, and find many different flowers,

but in the Ozarks their best sources are blackberries, wild sweet clover, persimmons, water willow, wild mint and wild fruit trees —plum, cherry and peach.

The nectar of the flowers is eighty percent or more water, and the sugars in the nectar are complex. To make honey, the bees have to evaporate the water and break down the sugars from complex to simple. When they gather the nectar, they suck it up through their long tongues and store it in a sac called a honey stomach. When this is full, they fly back to their hives and transfer the nectar to young house bees, who spread it, drop by drop, throughout the hive in the honeycombs. In the process of collecting the nectar, storing it in their bodies and transferring it to the house bees, the bees have added enzymes to the nectar which break down the complex sugars into simple ones, chiefly dextrose, levulose and sucrose.

The water in the nectar evaporates slowly from the droplets spread out through the hive, but the bees speed up the process by fanning with their wings, setting up currents of air from the hive entrance at the bottom to ventilation holes at the top.

On hot summer nights I like to walk out to the beehives back by the woodlot. At night all 60,000 bees are at home in each hive; most of them are fanning, and I can hear the hum of their wings long before I get near them. The air currents from the fanning of all those wings are so strong that when I stand in front of the hives in the dark I can feel a draft swirling around my ankles.

When most of the water is removed from the nectar, the bees cap each cell of finished honey with snowy white wax secreted in flakes from their own bodies. This finished honey has a very low moisture content, sixteen percent or so, dryer than air. This makes honey hygroscopic; it can pull moisture from the atmosphere. That is why I have to store it in tightly sealed containers once it is extracted; it is also the reason why baked goods made with honey stay moist and do not dry out the way they do when sugar is used.

When I harvest the honey and extract it at summer's end, I only take honey from combs that have been completely sealed with beeswax, honey that I know is ripe and will be flavorful. The uncapped honey, green and unfinished, will be thin and off-flavored, and I give it back to the bees. Last year, when I harvested 33,000 pounds of honey, at least 3,000 pounds of it was too green to extract, and I returned it to the bees. I am fussier about this than some bigger commercial beekeepers, but since I can't compete with them in price I try to produce very good honey instead.

In another week I'll have 6,000 pounds of honey processed and bottled; then I'll load up the truck with part of it and head out on the Interstates again to peddle honey and sleep in truck stops.

Paul is skilled and competent, and likes to do things well and neatly. During our life together, he assembled a large collection of tools, and when he left, he passed them on to me.

I set up a workshop for myself in the barn with the tools that I learned to use, and from time to time would try to sort through the others and the things that went with them. He had left coffee cans full of hardware labeled in his precise, draftsman's hand, "Three-Eighths Inch Bolts," "Springs of Assorted Sizes," "Insulated Staples," and the like. I would move the hardware from place to place and would pick up the tools, the purposes of which were unknown to me, and put them down again, depressed and defeated by my own ignorance, wondering yet again what I was doing trying to live out here by myself when I didn't even know how to use what might possibly be called a ratchet wrench.

After a long time, I piled all the tools that I did not understand on the upper two shelves of the tool cabinet, boldly scrawled below them "Mysterious Tools" and felt the better for it. Occasionally one or another of them loses its mystery and, comfortably useful, takes its place on my workbench. I had already sorted through the unlabeled boxes of hardware and filled coffee cans of my own, and marked them "Round Things," "Things That Fasten Other Things Together in Unusual Ways" and other categories that served me, if no one else.

Ermon came over and helped me sort through all the old Chevy parts in the barn, telling me which to save for repairs and which to scrap. He is a dark and moody man, a skilled mechanic who works on my truck and those of others when he has a mind to. He is brilliant and easily bored and turns down routine repair jobs. He is said to have a bad temper, but he has always been gentle and patient with me. He is said to be undependable, but I know better. Time after time he has come to rescue me when the Chevy has inexplicably ceased to run at an inconvenient time and in inconvenient places. He can always be counted on in a real emergency.

The day that he helped me sort through the Chevy parts we talked about the tractor sitting out in the shed. It was an old one, basically sound, but with many idiosyncracies and minor mechanical problems. When Paul lived here, he had used it to plow and disc the field where he tried to grow sweet clover for the bees. The thin, poor soil and drought defeated his efforts, and afterward he kept the field clear and the grass around the barn trimmed by brush-hogging several times a year. A brush hog is a heavy mowing machine pulled by a tractor.

An open field, if not cut frequently, is soon taken over by weeds, blackberries and multiflora roses. Years ago the multiflora rose, an exotic, was touted as a Miracle Plant. Its fruits are eaten by birds and its habit of thick, rapid, tangled and thorny growth could turn it into what the nursery catalogs used to call a Living Fence, one that even cattle will not go through. In the early 1940s the Missouri Conservation Department grew the plants and encouraged landowners to set them out. Birds did indeed like the fruit, but the seeds passed through their digestive systems fully fertile, and so birds planted new multiflora roses with their droppings. Instead of staying in tidy fence rows the multiflora rose spread rapidly over pastures where it was not wanted. It could not be dug out because the broken roots, hydralike, sprouted new plants. Recently strong herbicides have been developed that kill the rose, but they may contaminate water supplies and ponds. Brush-hogging multiflora roses at least keeps them from spreading.

I knew that I could not keep the tractor in good enough repair to do the brush-hogging; Ermon wanted it badly, but he could not afford to pay for it. So we made a deal. He could have the tractor if he would keep it in good working order and once a year bring it over and do the brush-hogging.

This worked well enough for several years. The tractor served him and he brush-hogged for me. But then he grew bored with the routine, and I had to nag to get the job done. Last year I nagged in vain.

I noticed that from an aesthetic standpoint I really preferred a field full of daisies, black-eyed Susans, chicory, Queen Anne's lace, wild pinks and scattered blossoms of sweet clover and alfalfa. Around and below the barn a lush stand of orchard grass and clover grew up untidily, but my eye became accustomed to it. I did fret about the multiflora roses, however. They had just been put on the official state noxious weed list, and most farmers were pouring on the herbicides and were paid for doing so by the Conservation Department, which was trying to make up for its past sins. I wondered if I should do the same.

Then I noticed something unusual. The growing tips of some of the roses were stunted, blighted-looking, twisted and unnaturally red. I did not know why until the extension agronomist explained that a new rose disease had showed up in our part of

Missouri. Not much was known about it, but this was what was killing back the growing tips on the roses in my field. State pathologists were studying it, but for now all they could tell us was that the disease was believed to be caused by a virus spread by mites. They suspected it would only affect weaker plants, but they had no idea what would really happen.

That the balance of nature should be restored right then and there while I was having trouble persuading balky Ermon to brush-hog seemed just a trifle too neat and pat, my own personal miracle for the control of the Miracle Plant. But although the timing may have been remarkable, what was happening was not. Multiflora roses are introduced exotics, and like other exotics such as English sparrows, kudzu and gypsy moths, they have spread aggressively, for they had no competition or checks to stop them. Now their very success has made them a dense and attractive habitat for the virus that has turned up.

I kept watching the multiflora roses in my field, and found that almost all of them were affected. None of them died of the disease, but they did not thrive either. I shall see what happens this coming year. Meanwhile I have stopped nagging Ermon. When he gets ready to brush-hog he will, and I will not ruin our friendship by harassing him. He tolerates my failings; I shall tolerate his.

Another pleasant thing has happened, a benefit from living in an untrimmed state. One night at the end of autumn I had been out to dinner with friends and returned home late. When I drove up in front of the barn the night was full of eyes. Eyes floating in the night, almond-shaped eyes everywhere, looking toward me, golden, gleaming eyes, eyes reflected in the headlights with no other body parts visible. Eyes surrounding me. Eyes. I turned off the headlights and quietly got out of the pickup. I was in the middle of a herd of deer. The night was moonless, but in the starlight their shapes were clear enough, and I could see them, no longer blinded by the headlights, relax and return to browsing in the thick, still-green orchard grass and clover that had grown up around and below the barn.

This winter is a harsh one. Ice, snow and bitter cold has made it hard for wild animals to find enough to eat. Each evening the deer have returned to feed. They have grown easy now, and feed close to the cabin. In the mornings I find their hoof prints and the bare patches of lawn where they have scraped away the snow to crop the grass.

I still do not know the use of the ratchet wrench. Paul was out here for a wary visit a few years ago and kindly explained its purpose to me, but what he told me has slipped my mind. I have concluded that I have been living a full and adequate life for fifty years without using a ratchet wrench, and so, no matter what its importance is to others, it is probably not important to me.

A touch of the Luddite, that, but I am learning to live with my own crotchets as well as those of my neighbors.

Perhaps the brush-hogging will get done this year; perhaps not. My eye has long ceased to find trimness pleasing, and as to the multiflora roses, I have never had much use for miracles, plant or otherwise. Besides, I want to see how virus and mite, two ordinary and nonmiraculous bits of life, will fit into the action taking place in my field. I have a ringside seat.

Once I tried to stop a war, and once I really did help start a labor union at a library where I worked. But, on the whole, the world has cheerfully and astutely resisted my attempts to save it. And now that I've spent my winter saving my particular ninety acres of it from the floodwaters of a dam, I am left to wonder, as usual, what I have done. Upon examination, the dam proposal turned out to be as lacking in reality as faerie gold, but the local people were sure that it was real, and so perhaps it was.

The controversy got its start at the end of the summer. I was harvesting the honey crop then, and immediately afterward started on my autumn round of sales trips, so I didn't hear much about it at first. But an item in the local newspaper explained that a Lakes and Dams Association had been formed to promote the damming of the river to create a recreational lake a few river miles downstream from my farm, just inside the strip of land along the banks of the river that the U.S. Park Service owns and supervises. The officers of the association were two men who work in the feed room at the general store. The news item amused me; it seemed a piece of folly that local people would laugh off.

But then, home from a sales trip a month later, it took me three hours to walk two blocks in town because people kept stopping me to ask what I thought about the dam. I didn't know anything about it, but I was given my questioner's full, colorful and highly charged opinion. In the café, there was talk of nothing else. A petition was being circulated asking the local conservative congressman to initiate a feasibility study for the dam's construction, and several thousand people had already signed it. I went back out on the road to sell honey, and when I returned again an anti-dam group had formed: Citizens for a Free Flowin' River. Aside from having a fine and ripply name, they caught my attention by issuing a map of the recreational lake which showed several portions of my farm under water. I promised myself that when I got off the road for the winter I would spend some time figuring out what was going on.

From the beginning, both dam and anti-dam were innocent of grubby reality. No one knew how dams got built or why. No trustworthy figures or facts were available from either side. But that was all right, because the argument took place over what was of value in this part of the Ozarks and expectations for the future —matters quite independent of reality, facts, figures or even a dam, for that matter. The people who wanted a dam were those who thought that it would be good to turn the town into the sort of place that had a McDonald's. Those who opposed it thought this would not be good at all. The dam was almost beside the point, and the eventual victory of the anti-dam forces was just a tiny rearguard action. People who want to exploit and change the Ozarks are still here, and will continue to suggest other plans for development.

Since 1909, proposals to dam this river have been made with varying degrees of seriousness. The two men in the feed room who had dusted off the plan came from families who had been involved in some of those earlier attempts. Once they started talking up a dam, others interested in development found reasons for it and the movement was born. The dam, it was said, could generate cheap electricity for the town. It would prevent damage from floods. Its construction would bring jobs to the area, and so would the businesses that a recreational lake would support. Land values would rise; indeed, property owners near the river immediately doubled the asking price for land they had for sale. The president of the local chamber of commerce was caught on television film, his eyes rolled heavenward, saying, "All I can see is dollar signs before my eyes."

But there was another compelling and contrary appeal that made people sign the petition for a dam. By inviting the U.S. Corps of Engineers to build a dam on U.S. Park Service land, the latter agency would get its comeuppance. It is hard for anyone accustomed to thinking of the popular U.S. Park Service as a benign bureaucracy to understand with what loathing the agency is regarded here. The local park was established in the mid-1960s to protect the river from attempts to dam it the decade before. Land along the river was bought up and removed from private to federal ownership; this shift in title to the river's banks has never been forgiven. As highways improved and it became easier for city people from Illinois and northern Missouri to get to the Ozarks, more people started floating the river, fishing in it and hiking along its banks, and the Park Service created new rules to regulate its use and protect the habitat. Local citizens, who remembered the old days when there were no rules and few tourists, were outraged, believing that their rights had been taken from them. They also resented, as an unwarranted constraint on the chance to make a buck, the Park Service limitation on the number of private concessionaires who rented canoes to tourists.

The dammers knew that they couldn't make the Park Service

go away, but they hoped they could punish it by withholding the river's waters behind a dam built by the U.S. Corps of Engineers, sensible folks, not like those fuzzy-minded park rangers who were always talking about the environment. The Corps certainly wouldn't issue a lot of silly rules that would keep local people from doing what they and their fathers and grandfathers before them had always done to and along the river.

The Free Flowin' Citizens, or at least the Ozarkers among them, didn't like the U.S. Park Service any better than did the dammers, but they didn't trust the U.S. Corps of Engineers either. They wanted the river to stay in private hands—namely, theirs, for many of them were landowners upriver. They did not like the idea of changing the town by bringing in even more tourists and outsiders than the federal park had already done. Many of them were concerned about the destruction a lake's waters could bring about to the wildlife habitat along the edge of the river.

For the past six weeks I have been spending my afternoons writing letters, telephoning and meeting in offices with appropriate government bureaucrats, elected officials and staffs of environmental organizations in order to find out how dams get built and what the impact of one would be on this area.

I discovered that this cave-riddled limestone is unsuitable for dam construction, that the Corps of Engineers builds flood-control dams only in urban and major agricultural areas, that costs of electricity generation have to be borne locally, that economic development in areas where dams have been built has disappointed local people and that rare and endangered plants in the state's Natural Areas System would be destroyed by the dam. I also found out that current practice requires large initial local funding, and that the federal government pays costs of a dam only when it can be justified for flood control—which this one could not. Other dams, such as those created for recreational lakes, would require local funding of at least half the cost. In a state still untouched by economic happiness, one which cannot afford to pay for its poor or its potholes, funding for a dam would be difficult.

I also found out that since the federal government had established the park in response to earlier efforts to dam the river, it had spelled out in the first sentence of the charter that the river must remain a "free-flowing stream." Those responsible for attempts to dam it in or above U.S. Park Service land would be taken to court by the Department of the Interior. Attempts to change the charter would not meet with favor in Congress at this time. In addition, the district Corps of Engineers, in response to local talk, had reviewed the earlier proposals to dam the river, and in light of current political and economic realities had concluded that a dam on the river was "unthinkable."

However, some people I talked to reminded me that the politics of the pork barrel is a queer business, and that economic policies change. I also learned that the Corps of Engineers had been forced by budget pressures to lay off personnel it could not justify, and might have an interest in keeping staffers busy with a dam proposal if there was enough public support for one.

So I met with the Free Flowin' Citizens and told them what I had learned. The information cheered them. They wrote letters to their state and federal representatives. They began holding forth at the café and on street corners. They printed up their own petition opposing the dam and began gathering signatures.

I asked a man I knew, a forceful public speaker who did not live on the river but who owned property on it that would have been flooded by a dam, if he would be the featured speaker at a public meeting the Free Flowin' Citizens wanted to hold. He agreed, and the meeting was held. Public opinion had already begun to shift away from support of a dam, and the meeting saw it disappear. The Dammers brought several advocates to speak at the meeting, but they were no match for the Anti-Dam speaker, a man who has made a career both public and private of his persuasive personality. He carried the day, explaining the Free Flowin' arguments in a winning manner; I had staged a bit of guerrilla theater, nostalgia for an old activist.

Ever since that meeting, the dam as a topic of conversation is greeted with embarrassed silence. The issue has disappeared as though it never was. In the café now they are talking about how hard the present cold snap is on the new calves being born in the pastures, and when I went in to buy feed today, the most fiery of the dam's advocates wanted to show me pictures of his new baby.

SPRING

I had missed the morning weather forecast so when I stopped in at the post office I asked my friend there if he had heard one.

"Supposed to be sunny today and tomorrow," he said, and then added, " 'Course I just heard that on one of them little bitty import radios, so's I don't know whether to believe it or not."

We were, all of us, rural mail carriers as much as anyone, looking for a good rain to settle the mud. This is mudtime. After a severe winter, when the ground was frozen a foot or more deep, the days suddenly have been warm and sunny and the ground has thawed quickly. The clay soil has yielded up its ice crystals to water, and in places on the back roads, which lack solid stone or gravel base, the clay has turned into something resembling chocolate pudding. We say that the bottom has gone out of the roads.

What we need is a good rain to settle the mud. I remember how strange that sounded the first year I lived here and heard it was what was required. I'm still not certain of all the chemistry involved, but a steady driving rain stabilizes the roads and turns the patches of chocolate pudding into a hard surface.

In mudtime the rural mail carriers have to skip part of their routes, and some people are as cut off as surely as if snowbound. I have been eager to get started on the spring beework, but I have to put it off until we have rain, for many of my beeyards are on the other sides of mudholes and I would not be able to drive through some of the pastures. Even my old Chevy pickup, "Press on Regardless," with a set of mud tires on the rear wheels and weight in back for traction, would not be able to get through, although, as pickups go, it is good in mud and snow, and cheerfully wallows through places that would mire down my three-quarter-ton truck.

Except for moving some hives last autumn, I haven't had anything to do with bees since the honey harvest at the end of summer. I had seen no weak hives that needed care during the harvest, and after that the snow asters had bloomed for so long and in such profusion that the bees made plenty of honey for themselves for the winter and did not need feeding.

Nobody but beekeepers seem to know about snow asters, and they call them by many names; frost asters, Michaelmas daisies, farewell summer, white rosemary and frostflower. *Aster ericoides* is the botanic name, and it describes the plant well. *Aster* means star and *ericoides* means with leaves like erica or heather. The plant, a bushy three feet or so in height, with tiny rayed blossoms like stars and fine, neat leaves, covers waste places throughout the Ozarks, blooming extravagantly from August until it is killed by a hard frost in October or November. But because snow asters are such common weeds no one but beekeepers and bees get excited about them. Beekeepers think snow asters are beautiful. The blossoms secrete a strongly scented nectar during their flowering months, and the bees gather it in such quantity that their hives reek with the odor. I have never tasted the honey, but it is dark and supposed to be strong. Because of weather conditions, the aster bloom occasionally fails, and then, even though I am very conservative about the amount of honey I take from the bees in August, I have to feed them in the autumn to get them through the winter. But last year the hives were heavy with aster honey by October, seventy-five pounds or more, and I did not have to feed.

As the weather becomes cooler in the autumn, the bees chink up every crack in their hives with propolis, a gummy resinous material they gather from buds and bark to seal out the wind and cold. If there is anything a bee hates, it is a draft. Opening a hive

in the late autumn or winter breaks those seals and can harm the bees, so I try to leave them alone as much as possible in cold weather.

It has been a long time since I have been with the bees, and I miss them and wonder how they are doing. Yesterday afternoon

I took a cup of coffee out behind the barn to a spot protected from the wind and drank it in the sunshine. The bees discovered me there. One settled on the back of my hand and walked daintily along my fingers to inspect the contents of my cup. Finding it unsatisfactory, she flew away. The bees are active and flying and ready to get on with the new year, and so am I. But not today and not tomorrow. Not in mudtime.

Yesterday afternoon I walked up to get the mail in rubber boots which grew heavy with caked mud before I had walked a quarter of a mile. When I reached the box I could see a truck mired in the mudhole just beyond it. Its owner, Bob, a man who lives on the other side of town, was spattered with thawed cold mud and sorely out of temper. Those of us who drive the road regularly know a detour around it, but Bob was unfamiliar with the road and had tried to drive his one-ton flatbed truck through it. By the time I got there very little of the rear wheels could still be seen, and the mud was up over the differential. A couple of neighbors and the mailman were also there helping. One neighbor with a jeep had tried to pull out Bob's truck, but the powerful gooey mud and the weight of the truck had defeated his attempt.

Another neighbor had brought a shovel, and when I got there Bob had just started to dig. Everyone thought that if he could dig down under the differential we could pile enough rocks under it to give footing for a jack, and could then jack the truck up out of the mudhole. I helped gather rocks from the edge of the pasture next to the road. I had just dropped a load into one part of the hole when Bob shifted position to start digging in a new spot and his boot refused to follow his foot. The mud had seized it, and he stood poised, one bare pink foot in the air and a look of indecision on his face.

"She-ittttFARR!!!!" he bellowed.

"Shitfire" is an Ozark expletive, extraordinarily relieving and satisfying if correctly pronounced.

Bob grinned and squelched his naked foot down in the cold mud. We rock carriers cheered.

After a while we had enough stones in place to give a base for the jack, and Bob retrieved his mud-filled boot. He jacked up the truck, freed the differential and the neighbor's jeep was able to pull him out.

The mailman handed me my mail. I handed him back an advertising flyer and he used it to wipe clean the mud from his hands. Then I walked home.

There was a day last winter when I badly needed springtime, and since it did not appear that spring was going to come to me I went to it. One of the good things about living here is that the lay of the land between river and creek and the steepness of the terrain create different climates and seasons all over the place.

No matter what the calendar says, a few sunbeams bring springtime to the south face of the creek hollow, and so on that sunny winter day I went there. I bundled up in insulated coveralls, insulated boots, scarf, woolen hat and two pairs of gloves, and set off with the dogs across the snow-covered field toward the southeast to the high rocky point where the creek and the river join.

Overhead a rough-legged hawk was quartering the field, hunting in vain for mice. By the calendar's springtime he would be in Canada on the way to his northern breeding grounds; so would the golden-crowned kinglets I discovered at the edge of the field in the cedar and pine trees. Tiny, cheerful, gray-green birds with a patch of brilliant yellow on the top of their heads, the flock of kinglets were inspecting twigs for insect eggs with such keenness that I bothered them not at all as I stood and watched. The cold began to seep in at my feet, however, so I walked on through the woods, breaking a path in the crusty snow.

The soil is poorer and thinner here, on the narrowing strip of land sloping southward. Hardwood trees give way to scrub, grasses and, finally, at the rocky cliffs several hundred feet above the juncture, to the modest plants of a limestone glade. The view from this rocky point is spectacular, and the walk to it is one of my favorites in the winter. With the leaves off the trees, the structure of cliffs along the river and creek and the hills beyond are visible—all the foundations of this beautiful land are there to see.

Water melting from the snow above oozed through the low grasses and dripped from the cliff edge; there, in the sunshine, mosses and lichens were holding their own private but exuberant

springtime. All were shades of tender, sweet and vibrant green, and several varieties of mosses were covered with fruiting bodies.

I was warm out of the wind, so I pulled off my heavy coveralls and scrambled down the cliffs to look for the blossoms of harbinger-of-spring which should be growing in the rich soil at the creek's edge. *Erigenia bulbosa* takes its genus name from the Greek *ery-geneia,* "early born," Homer's epithet for Eos, goddess of the dawn. Poking in the leaf mold at the foot of the cliffs, I found the early born, with its bulbous root and clusters of small white flowers.

The climb back up the steep cliffs was harder than my sliding descent, and I had to stop once to catch my breath and peel off my sweatshirt. At the top I found a sunny niche among the lichens and moss and sat down, taking off my heavy boots and socks. Across the creek, the snowy, shadowed woods on the north face of the hollow insisted that it was still winter.

That was in January; now, in April, it is springtime everywhere. The kinglets and the rough-legged hawks are gone. The woods are full of migrating warblers, and the hummingbirds have returned to the feeder by the windows. There are wildflowers everywhere, even on the north face of the river's gorge, on Pigeon Hawk Bluff, where spring comes last and most grudgingly of all. This slope is frosted with the feathery white blossoms of the small tree that as a child in Michigan I had known as shadblow. Ozarkers call it sarviceberry. The tree's botanic name is *Amelanchier arborea.* On the ground under the trees are the pink and white blossoms of rue anemone and the white-blossomed, liver-leafed hepaticas.

In the grass around the cabin the bees are eagerly working golden dandelions and ignoring the bluets and violets. The violets, purple, blue and white, are growing in such profusion that the air is scented with their fragrance. From the upland woods, the sweet odor of wild plum blossoms comes in on every breeze. The bees like wild plum blossoms, and so do I. They smell exactly like cherry Lifesavers taste.

The violets bloom along the dirt roads I drive down to get

to my beeyards, and in one yard a special violet, the Johnny-jump-up, which looks like a little pansy, has taken over. I had fenced off these beehives, so that the cows cannot graze near the hives, and the Johnny-jump-ups have pushed out the pasture grasses and surrounded the hives with blossoms, blue and purple with yellow and white centers.

I have a friend, an amateur botanist, who carries a sketch pad with her and draws each flower she identifies. There is no surer way, she tells me, of learning a plant, simply because of the painstaking observation needed to put it on paper. She is right, and some day when I have fewer bees and more time I would like to do this too. But now I am so occupied with the bees that I have not even had time to walk back to the point at the juncture of the hollow and the river where I first found springtime. The season is well advanced there, and I know what I should find, for I have seen it in other years: the wildflower whose common name I like best of all, hoary puccoon. Puccoon is an Indian word. I do not know its meaning, but I like its sound. There are other puccoons, hairy and narrow-leaved, but mine, the hoary, has thick clusters of yellowish-orange blossoms. They are as pretty and showy as any cultivated flowers in a formal garden, and make the rocky glade overlooking the river and creek beautiful beyond telling.

This afternoon when I got back from the beeyards, Nancy was waiting for me. It was Thursday, she pointed out. Nancy works in the office of a small factory in town, and because she is competent she takes her job seriously and finds it hard to leave it behind when she locks the office door. Once, a few years ago, I was listening to her complaints about office politics, and pointed out to her that she was always in a bad mood by Thursday. I could remember what Thursday was like in an office. Friday is still ahead, and by Thursday work deadlines are overwhelming, co-workers have called in sick, and the stupidity of the boss and one's own clear crystalline good sense seem in sharpest contrast. I told her that a good thing about working in an Ivy League university had been that on Thursday there was always a sherry hour to be found somewhere on campus to help put things back in perspective.

Nancy has not lived in university circles and did not know about sherry hours, so I introduced her to them. I now keep a bottle of sherry on hand. Word has gone round, and sometimes we are joined by others for a late-afternoon Thursday sherry hour. But today she was the only one waiting for me, and we decided to walk down to the river first. She wanted to tell me about the really dumb thing the new vendor had done, the terrible mixup with an invoice, what her supervisor had said to her, what she had said to the supervisor and how irritating it was when the computer broke down for the third time.

I nodded and said Umn. I remember what it was like to work in an office, particularly in the springtime.

Before we got to the river, we found patches of Dutchman's-breeches, fern-leafed plants with pinkish-white blossoms shaped like a pair of tiny pantaloons. Nancy kneeled down to take a better look. The river's banks were covered with *Mertensia virginica,* one of the many wildflowers that are called bluebells. The clusters of sky-blue, bell-like flowers were growing so thickly that we could not walk among them without crushing them, so we simply stood and admired.

Back at the cabin we poured glasses of sherry and took lawn chairs to the deck of the barn loft. It is high up there, so we could still have sunshine.

We drank a toast to springtime. The low rays of the sun spread golden light across the greening fields. The breeze was fragrant with wild plum blossoms and violets. There was no more talk of invoices. Instead we sat silently, sipping sherry and watching the buds swell in the western woods until the sun went down behind them.

Spring 187

———————

Today's mail has a notice in it that my first spring shipment of queen bees will be here next Wednesday. I have them shipped each April in batches of twenty-five and there is always a lot of excitement in the post office when they arrive.

They come by airmail from a honeybee breeder in Georgia, each queen in her own small wood-and-screen-wire cage with three or four attendant worker bees to feed and take care of her; she cannot take care of herself, for she is a specialist, an egg layer and nothing more. Each cage has a plug of sugar candy at the end, which the bees use for feed. If the weather is too bad for me to put the queens out in the hives, I must coat the screen wire with water so that they will have something to drink, too.

The cages are banded together, and the bees can be heard buzzing in a disturbed-sounding way inside; my mailman always calls me promptly when they come in, for he believes that they will somehow get out and he is a little afraid of them.

After I pick them up from the post office, I bring them home and pry apart the individual shipping cages to see that each queen, long and elegant, is alive. (The bee breeder will replace any that arrive dead.) I buy ordinary Italian queen bees, the usual commercial strain, but even these cost seven dollars each, so I can't afford to lose any. If I bought hybrid queens, as many amateur beekeepers do, the cost would be higher, and an artificially inseminated queen would be thirty-five dollars or more. After I check each queen in her cage, I put the cage down carefully on the kitchen table. Queen bees are jealous, and would kill one another if they could. Sensing the presence of the rest of the queens, they shriek in challenging, high-pitched voices—*ze-eee-eep, ze-eee-eep, ze-eee-eeep.* There is murder in their hearts.

If left to their own devices, the bees in my hives raise a new queen for themselves every few years, when the one that they have becomes too old to lay fertile eggs. To raise a new queen, the worker bees—females with atrophied sexual capacities—select a freshly laid fertile egg, one that would develop into a sister worker bee, and feed the larva that develops from it royal jelly, a glandular secretion of their own making that is rich in B vitamins. It is this food and this food alone that creates a queen bee, who will emerge from pupation less than two weeks after the egg is laid.

During most of their lives queen bees shun the light; they run and hide when a hive is opened. But the newly emerged queen is a virgin and is attracted toward light. Urged on by the worker bees, she flies out of the hive high up in the air on her mating flight. Surrounded by drones—male bees—she will mate ten or more times serially, securing a lifetime supply of spermatozoa which she will use to fertilize the eggs she will lay. In mating, a drone everts his penis into the queen's sting chamber, where it remains as the drone pulls away and falls lifeless to the ground, his function in the bee colony completed. Drones are not physiologically suited to forage for nectar or pollen, nor do they have stingers with which to defend the colony. Their role is solely to mate with a nubile queen. After the spring and early summer mating season, drones are seldom to be found in a colony of bees; the workers bar them from the hive, and without its food stores they starve. A colony of bees with a failing queen will tolerate

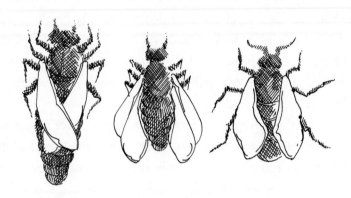

drones for a much longer time, so that they will be available for mating. Drones are large, stout, furry bees with huge eyes and are easy to distinguish at a glance. Their obvious presence late in the summer or during the fall means that a colony has queen problems, is in trouble and needs checking.

I let most of my colonies of bees raise their own queens, believing that they are a better judge of when they need a new queen than I am. The strains of bees that develop in such manner are best suited for this area and this climate. I keep records on each hive, and as long as it produces an average amount of honey or better, I let the bees go their own way. But when a hive ceases to be productive I investigate, and if it appears that the queen is a poor one, I mark the colony for re-queening during the following spring.

Bees' aggressiveness, their quickness to sting, is a genetic characteristic, and often beekeepers, especially those with only a few hives in the backyard, re-queen colonies that have become aggressive. But I never re-queen a colony that makes lots of honey just because the bees are fast to sting me. They know how to make honey, and for that I am grateful; I will be extra gentle with them, use a bit more smoke from my bee smoker to quiet them when I open the hive and accept a few more stings than usual.

I also use the queen bees I buy to start new replacement colonies for ones that have died out during the previous year. To start a new hive or to re-queen one, I follow the same procedure. I take three frames, or honeycombs, that are filled with eggs, larvae (called brood), and young bees, and put them in a single-story empty hive. Then I add two more frames filled with honey and four empty frames of honeycomb. In between two of the frames of brood in the center, I wedge the new queen in her cage, after poking a little hole through the sugar plug at the end. If I let the queen out of her cage and put her into the hive directly, the bees on the brood frames would kill her, for they still consider themselves loyal to the queen from whose hive I have taken them. By leaving her in her cage, they have a chance to become ac-quainted with her through the screen wire which protects her. Gradually the memory of their old queen fades, and by the time they have chewed through the sugar plug at the end of her cage, they are ready to accept her as their own. In warm, fair weather the process takes a day or two. If the new queen is a good one, she immediately starts to lay eggs and the single-story hive is then ready to take to a permanent location or to use to re-queen a poor hive. I do this by finding the old queen and killing her, putting the new nucleus hive in the place of the old one, and adding the old bees to the new.

Spring beework requires time, patience, some skill and a strong back. It also requires a clear mind and concentration. There is nothing that so focuses the attention as opening a hive of bees. At full summertime strength, a bee colony has about 60,000 bees in it, and in the springtime half or three quarters of that. When I open a hive, the sheer number of bees fussily tending to their business makes me tend to mine, which is their care.

In recent years, because more and more of the honey market is being taken over by cheap imported honey, there are fewer U.S. commercial honey producers than there used to be. At last count, I saw we were down to twelve hundred, which is four hundred fewer than six years ago. A day is coming—probably quite soon—when I may have to admit that in a price-conscious market I can't meet the competition of South American honey, and will have to make a living some other way.

But after keeping bees, whatever will I do?

My bees cover one thousand square miles of land that I do not own in their foraging flights, flying from flower to flower for which I pay no rent, stealing nectar but pollinating plants in return. It is an unruly, benign kind of agriculture, and making a living by it has such a wild, anarchistic, raffish appeal that it unsuits me for any other, except possibly robbing banks.

Then there is that other appeal, the stronger one, of spending, during certain parts of the year, a ten- or twelve-hour working day with bees, which are, when all is said and done, simply a bunch of bugs. But spending my days in close and intimate contact with creatures who are structured so differently from humans, and who get on with life in such a different way, is like being a visitor in an alien but ineffably engaging world.

In town I am known as the Bee Lady. Whatever could I do to equal that?

Hoohoo-hoohoo . . . hoohoo-hoohooaww. My neighbor across the river is doing his barred owl imitation in hopes of rousing a turkey from the roost. It is turkey-hunting season, and at dawn the hunters are trying to outwit wild turkeys and I listen to them as I drink my coffee under the oak trees.

Hoohoo-hoohoo . . . hoohoo-hoohooawww.

GahgahGAHgah replies an imitation turkey from another direction. I know that neighbor, too. Yesterday he showed me the hand-held wooden box with which he made the noise that is supposed to sound like a turkey cock gobbling. It doesn't. After the turkey cocks are down from their roosts, the hunters, by imitating hen turkeys, try to call them close enough to shoot them. The barred owl across the river once showed me his turkey caller. He held it in his mouth and made a soft clucking noise with it.

"Now this is the really sexy one," he said, arching one eyebrow, *"Putput . . . putterputput."*

It is past dawn now, and I imagine both men are exasperated. I have not heard one real turkey yet this morning. The hunting season is set by the calendar but the turkeys breed by the weather, and the spring has been so wet and cold that their mating has been delayed this year. In the last few mornings I have started hearing turkeys gobbling occasionally, and it will be another week or two before a wise and wary turkey cock could be fooled by a man with a caller.

There are other birds out there this morning. The indigo buntings, who will be the first birds to sing in the dawn later on, have not yet returned to the Ozarks, but I can hear cardinals and Carolina chickadees. They wintered here, but today their songs are of springtime. There are chipping sparrows above me in the oak trees and field sparrows nearby. There are warblers, too; some of their songs are familiar, and others, those of the migrators, are not. I hear one of the most beautiful of birdsongs, that of the white-throated sparrow. He is supposed to sing "Old Sam Peabody, Peabody, Peabody." This is the cadence, to be sure, but it gives no hint of the lyrical clarity and sweetness of the descending notes of his song.

I slept outdoors last night because I could not bear to go in. The cabin, which only last winter seemed cozy and inviting, has begun to seem stuffy and limiting, so I spread a piece of plastic on the ground to keep off the damp, put my sleeping bag on it and dropped off to sleep watching the stars. Tazzie likes to be near me, and with me on the ground she could press right up to my back. But Andy is a conservative dog who worries a lot, and he thought it was unsound to sleep outside where there might be snakes and beetles. He whined uneasily as I settled in, and once during the night he woke me up, nuzzling me and whimpering, begging to be allowed to go inside to his rug. I think he may be more domesticated than I am. I wonder if I am becoming feral. Wild things and wild places pull me more strongly than they did a few years ago, and domesticity, dusting and cookery interest me not at all.

Sometimes I wonder where we older women fit into the social scheme of things once nest building has lost its charm. A generation ago Margaret Mead, who had a good enough personal answer to this question, wondered the same thing, and pointed out that in other times and other cultures we have had a role.

There are so many of us that it is tempting to think of us as a class. We are past our reproductive years. Men don't want us; they prefer younger women. It makes good biological sense for males to be attracted to females who are at an earlier point in their breeding years and who still want to build nests, and if that leaves us no longer able to lose ourselves in the pleasures and closeness of pairing, well, we have gained our Selves. We have another valuable thing, too. We have Time, or at least the awareness of it. We have lived long enough and seen enough to understand in a more than intellectual way that we will die, and so we have learned to live as though we are mortal, making our decisions with care and thought because we will not be able to make them again. Time for us will have an end; it is precious, and we have learned its value.

Yes, there are many of us, but we are all so different that I am uncomfortable with a sociobiological analysis, and I suspect that, as with Margaret Mead, the solution is a personal and individual one. Because our culture has assigned us no real role, we can make up our own. It is a good time to be a grown-up woman with individuality, strength and crotchets. We are wonderfully free. We live long. Our children are the independent adults we helped them to become, and though they may still want our love they do not need our care. Social rules are so flexible today that nothing we do is shocking. There are no political barriers to us anymore. Provided we stay healthy and can support ourselves, we can do anything, have anything and spend our talents any way that we please.

Hoohoo-hoohoo . . . hoohoo-hooaww.

The sun is up now, and it is too late for a barred owl. I know that man across the river, and I know he must be getting cross. He is probably sitting on a damp log, his feet and legs cold and cramped from keeping still. I also know the other hunter, the one with the wooden turkey caller. This week what both men want is a dead turkey.

I want a turkey too, but I want mine alive, and in a week I'll have my wish, hearing them gobbling at dawn. I want more, however. I want indigo buntings singing their couplets when I wake in the morning. I want to read *Joseph and His Brothers* again. I want oak leaves and dogwood blossoms and fireflies. I want to know how the land lies up Coon Hollow. I want Asher to find out what happens to moth-ear mites in the winter. I want to show Liddy and Brian the big rocks down in the creek hollow. I want to know much more about grand-daddy-longlegs. I want to write a novel. I want to go swimming naked in the hot sun down at the river.

That is why I have stopped sleeping inside. A house is too small, too confining. I want the whole world, and the stars too.

Last winter a bird never before seen in Missouri or any of the lower forty-eight states, the slaty-backed gull, a native of the Bering Sea, was blown into the state on an arctic cold front and was spotted near the Mississippi River by a birder experienced enough to realize that this rather undistinguished-looking gull was different. "When I first saw it, I had no idea what it was, but I knew what it wasn't," said Bill Rudden, who holds several state titles for rare-bird spotting.

That there can be titles for rare-bird spotting is an indication of what has happened to bird watching now that it has become birding. It is a competitive affair. Field days are held in which one demon birder pits his skills against another, and the winner is the one who has sighted the most species of birds within a twenty-four hour period, aided by tapes of the birds' song, which entice them into view. Birders keep life lists of birds seen, and travel around the country for the single purpose of adding to their lists. It is a trophy approach to natural history. I once asked a ferocious birder about a spring warbler that I had seen in my woodlot during migration time. It had puzzled me, and I was not sure that I had seen what I thought I had seen. Did he know the warbler? Had he ever seen it? "I don't know," he replied, "I'd have to check my life list."

The slaty-backed gull's appearance in Missouri was an event for competitive birders, and they flocked, pardon the expression, to the ice floes on the Mississippi River, pencil in hand, to be able to add the bird to their lists. The confused gull had flown into the state on wing power buoyed by arctic winds, but the Audubon Society flew its top muffler-wrapped birders in on an airplane. After positively identifying the gull, they called a news conference, and the gull's picture and story appeared on the front page of the New York *Times*. The slaty-backed gull was duly noted to be the 378th bird species to be identified in Missouri, and was solemnly declared to be the state's Best Bird.

I've always disliked making lists and taking part in organized fun and games, and the few birding expeditions that I have been on have only reinforced this dislike. I will never make a birder.

That said, I feel free to award my own title of Best Bird wherever I please, and to change it from time to time. This week I have given the title, in its plural form, to a pair of cowbirds. Birders wouldn't give a cowbird a look. Black birds with brown heads, they are far too common, and appear at one time of the year or another in every part of the country, sometimes flocking in numbers, often foraging among cattle for insects disturbed by the animals' hooves. Their habit of laying eggs in the nests of other birds for the latter to raise, to the detriment of their own young, has, it is claimed, reduced the population of orioles, tanagers, warblers and vireos. As a result they are considered to be rather trashy birds. One of my bird books refers to them as "utterly irresponsible," and goes on: "In keeping with its unholy life and character, the cowbird's ordinary note is a gurgling, rasping whistle, followed by a few sharp notes." My other bird books speak no better of their song. But these books are wrong, which is why cowbirds are my Best Birds this week. Two of them have been courting, and when I first heard them I was inside the cabin and came out with my binoculars to see what was singing such a beautiful song, reminiscent but slightly different from the liquid, bubbling trill of red-wing blackbirds. It was the cowbirds.

Since I became aware of this song, I have been watching the cowbirds all week. They are conducting their love affair in the treetops and on the power lines that run between the barn and the cabin, so I can set up a lawn chair in the space between and watch them comfortably through my binoculars. The male throws its head straight back, fluffing up all his feathers elaborately and singing a piercingly sweet song. Wing-fanning and tail-wagging follow, and the display culminates in a plebian-sounding *s-k-r-a-a-a-t,* a finale that the female seems to find impressive. I'm delighted to have heard and watched their carryings-on, and to report that I think their courting will be successful. I hope everything works out for the two of them.

Last week my Best Birds were a pair of Baltimore orioles. Even trophy birders would concede that these birds are beautiful, although they probably would insist that they are far too common to be of interest, since they are summertime residents all over the United States. The males are brilliantly colored, orange with black hoods and backs; the females are somewhat duller. They eat insects and caterpillars and are usually seen feeding in trees, so I was startled when I walked through the living room last week and saw a flash of orange near the syrup-filled hummingbird feeder that hangs in front of the window—a dainty bit of glass and red plastic with four slim wells. I waited for whatever it was I had seen to reappear, and in a few minutes a pair of Baltimore orioles returned. The orioles are robin-sized, and clumsily found toeholds on the feeder, tipping it from side to side to spill out drops of the syrup, which they dabbed up with beaks more suited to an insect diet. I could hardly believe what I was seeing, but over the next several days I watched them repeat their perform-ance, clutching awkwardly at the wildly swinging feeder while they slurped up syrup, offending the hummingbirds and driving them away.

Then, at the end of the week, the newspaper from the county seat, some thirty miles away, ran a front-page picture story of a Baltimore oriole feeding at a town resident's hummingbird

feeder. The woman who owned the feeder had spotted an oriole there the same day I had first seen my pair. She telephoned the state conservation department's local office for an explanation of this bizarre behavior and was dismissed by the agent, who assured her that a Baltimore oriole would not be at a hummingbird feeder and that she had misidentified the bird. Annoyed, she telephoned the newspaper office, which sent out a photographer who took a series of pictures showing an oriole at his gymnastics.

What had happened? Did Baltimore orioles all over the Ozarks just last week independently discover that those red plastic feeders had good stuff inside? Or did word go out that, though not as tasty as caterpillars, those red things were worthwhile if you tipped them just right? If so, how was this information passed around? I don't know the answers to those questions, but thinking about them beats the making of bird lists.

My idea of bird watching is to stuff my pockets full of bird books, hang my binoculars around my neck and go out with the dogs to sit quietly under a tree by the edge of the woods where there is both open and leafy habitat. My walk to the spot will have been noted by all the birds around, for their eyesight is keener than mine. For a time I'll neither see nor hear a bird. Andy will trot off to track rabbits, but Tazzie will sit beside me as close as she can get and sigh. She knows bird watching is a quiet and solitary business, and that if she is to be allowed to stay she must be still. In half an hour the birds are no longer alarmed and resume their usual affairs—and I am privileged to watch them, sometimes so close that I do not need my binoculars.

Even better than sitting in the woods is to go out to the rope hammock hanging between two pine trees below the cabin. I crawl into it with my books and binoculars late in the afternoon when I am tired and when the birds are feeding for the last time before dusk. Just to one side of the pine trees is a patch of sumac and other low, wild, shrubby growth, and beyond that is the open field. Birds congregate all about me—common enough birds, to be sure: goldfinches, indigo buntings, hummingbirds, blue-winged warblers, prairie warblers, yellow-breasted chats and all the rest of the warm-weather residents whose names, songs and behavior I have become acquainted with over the years. They are like old friends to me. I enjoy seeing what they are up to, and listening to their end-of-the-day songs.

Spring 202

✿

Early this morning, before I left to work with the bees, I went to see how the rhubarb is coming along. My brother, Bil, and his wife, Ann, are going to be here for a visit soon and I want to make them a rhubarb pie. It is Bil's favorite.

Back when there were two of us here to eat and do the work, I had a big garden between a grove of pepperidge trees and the old road down to the river. I enjoyed the work that was necessary to produce, by midsummer, the neat rows of beans and beets, carrots, potatoes, corn, tomatoes, Swiss chard, onions and lettuce of good flavor and amazing hues. I mulched the plants with newspaper and straw to keep moisture in and weeds out. I planted peas along the inside of the garden fence; the plants blossomed and produced tender pods. And throughout the garden I had bright gold and orange marigolds. They helped to protect the vegetables, because bugs dislike and avoid their flowers, avoiding the vegetables at the same time.

For one person a garden is not worth the work, which comes, for the most part, just when I am busiest with the bees. But sometimes I still clear away a little corner of the patch and plant a few tomatoes and lettuces for summer salads. This year, however, there are only the tough perennials that can hold their own against weeds. A few strawberries still grow, but the turtles will get the fruit before I do. Rhubarb, asparagus and feisty elephant garlic are flourishing. The herbs, particularly mint, feral plants that they are, are spreading. The garden is being taken over by

wildness. Shoots of pepperidge growing up in it from the roots of a nearby tree will become a new grove in a few years and claim the garden as their own.

Living even the modestly domestic life that I do in a wild place requires a constant balancing act. While I was out in the garden looking at the rhubarb, wild things were asserting their rights in the cabin. The termites were swarming up out of the floor in the old part which is my living room. They have done this for a ten-to twelve-hour period in the springtime for the past several years, and today was their day. The living room is the original one-room cabin built here fifty years ago. Like many Ozark cabins, its underpinnings are a series of two-by-fours balanced on paired rocks on the bare ground. It is protected, damp and warm under there, a perfect habitat for termites, and they feast upon the two-by-fours and probably on the floor itself.

Our houses are only a minor source of food for termites, and if we humans did not have a penchant for building things out of wood we would look upon them with kindness for the role they play in turning dead and fallen wood into soil. The soil provides the base and nutrient for new-growing plants, and so the termites give the wheel of process an important turn.

There are more than two thousand species of termites, but in North America, not a favorable climate for them, there are a scant forty-one. Of these, the subterranean termites who belong to the family Rhinotermitidae are the ones who most alarm human beings.

Termites are social. Their eating habits make them so. Although some people call them flying ants, they are not related to ants but to cockroaches. Like some roaches, they contain within their gut microorganisms that process cellulose, transforming it into a food they pass between one termite and another by anal feeding. This makes social organization necessary.

Most of the termites within a colony are sterile workers, and, in some species, "soldiers," with big heads and jaws, who protect the colony against enemies, usually ants. The workers of these subterranean species do not have the hard chitinous body covers of many other insects. They must construct tunnels to their above-ground food sources in order to protect their pale, delicate, wingless bodies. The tunnels are made of fecal pellets, a building material that also requires a cooperative social structure. These tunnels, scaling the face of a concrete foundation, are often the first sign we have that our houses are providing termite dinners.

The termites I saw swarming out of my floor this morning were the reproductives, the alates, winged, unmated males and females who had emerged from the underground nests to establish new colonies. Their wings are weak and they cannot fly far. When they land they break off their wings by pressing the tips to the floor. After a minimal courtship the future king and queen of a new colony pair off, select a suitable spot for a nest and begin excavating a burrow from which they will never again emerge. From the eggs that the queen lays, sterile offspring will appear to build up the colony and do its work. The mother queen lives for many years, and when she begins to fail she is replaced by secondary reproductives, so it is possible for a termite colony to live as long as its food supply.

The termites that eat our North American houses, members of the genus *Reticulitermes,* are regarded as fairly primitive species by entomologists. The tunnels that they build to their food supply free them only a little from their food sources. The most primitive of all are prisoners to their food and must live within it, a happy circumstance for those who want to study them, like the researcher I read about who for years kept a prospering colony of primitive termites in a block of wood inside a bottle in his office. No special heat or humidity was required, and they needed no care or attention.

The majority of the world's termites, however, belong to more evolutionarily advanced species than those minimalists, or even those who nibble away at our houses. They are found in tropical climates, and are diverse in form, social organization and habits. The most advanced are found in the Old World tropics and have managed to free themselves completely from dependence on existing wood. They raise their own food inside mounded nests on the ground, and build round combs in special chambers within their nests; on these they grow fungus gardens that serve as a food supply. Big populations are needed to build and maintain these complex nests; approximately two million termites live within each of the huge mounds.

I don't suppose there are that many of these less-sophisticated termites under my floor, but even if there were I wouldn't worry about them. Perhaps this is because I have never owned a television set and am therefore spared what I hear are terrifying advertisements that depict termites and scare homeowners into buying the services of pest-control companies. But I also know that in temperate climates it takes termites a long time to eat up something as big as my cabin. I still have time before wildness takes it, and tonight I will simply sweep up the shed termite wings on the living-room floor.

I think it is sensible to let the garden go; I have neither the need for it nor the time to work it. But I am not yet quite ready to give up on shelter, so I am planning to replace some of the cabin's sills and studs with treated lumber, to tear out the living-room floor and put down fieldstone. This will not only be discouraging to termites, but will provide enough mass to equalize the temperature in the room by storing heat from the sunshine that streams through the windows. I cannot make this change right away, because the labor and materials are too expensive. But termites who live in a temperate climate are leisurely creatures, and I believe I shall have the money for a new floor before they have finished with the old one.

A young friend of mine, Winnie, who lives in Boston, attended a conference in Kansas City recently. After her meetings were finished, I drove there to pick her up and brought her back for a visit, which has just ended. We hiked, read aloud from a book she had brought, talked and sunbathed down by the creek. One day we went canoeing on the river and saw great blue herons, a snake eating a fish on a rock and the yellow and scarlet blossoms of columbine clinging to cliff faces everywhere.

Another day we went into town to do errands; when we were done, Winnie suggested we have a drink together. It is the sort of thing we do when I visit her in Boston; it is harder here in a dry town of one thousand three hundred ostensible teetotalers.

There is one bar outside the city limits, down by the sewage lagoon. It is dark, forbidding-looking, dreadful. A couple of years ago a man who had been drinking and shooting pool there picked several fights in succession, went out to his pickup and returned to settle matters with his chain saw running at full throttle. The crowd in the bar was badly frightened: a pool cue was sawed in two and several people cut up before the man was stopped with a charge in his stomach from a 20-gauge shotgun another man had handy. It is not the sort of place where Winnie and I could enjoy a quiet drink.

But when she proposed it, I remembered a restaurant, also outside of town, that had recently opened and had, tentatively and with considerable daring, advertised a Happy Hour on Friday afternoons. It was a Friday, and Winnie thought Happy Hour in the Ozarks was not to be missed, so we drove out there. Inside we found the town yuppies, a small group who cling together with the pathetic tenacity of the pharmacist, doctor, government inspector and schoolmaster in Russian novels about provincial life.

I introduced Winnie, and we sat down. She looks younger than her twenty-six years, and when the waitress took her order she asked for proof of age. Winnie handed over her Massachusetts driver's license, but the waitress shook her head. Not good enough; for a driver's license to be proof, it would have to be from Missouri or an adjacent state.

Winnie looked disbelieving, but this is the Show Me state, and those distant places, if they do indeed exist, might not know how to get their important facts straight.

Winnie fumbled through her billfold, slightly embarrassed. "What about my Lord and Taylor charge card?" she joked.

The waitress looked with awe at the piece of green plastic.

"I'll take it to the manager and see," she said to our surprise. She returned a few minutes later. "He said it was acceptable," she reported solemnly. If you can handle the responsibility of a Lord and Taylor charge card, then, by golly, you can drink a margarita in Missouri.

The yuppies were impressed with the power of big-city ways and asked Winnie all about Boston. She, in turn, liked them, and asked them about their jobs and lives. They drank their margaritas, told lies and had a good civilized time.

I have a lot of visitors in the spring and summer. Most of them I know and love, like Winnie, but over the years I have more and more visits from people I do *not* know. They are people who have grown weary of cities and who want to move here to live the Simple Life. Sometimes they would like me to tell them how to make a killing in bee farming; some are just looking for an idyll, a bucolic Simple Life, never mind the bees—though of course they would like a hive or two.

I find the hopeful bee-agribusiness visitors touching and appealing; many of them are young and they bring out the mother in me. They are often working at dull jobs they do not like, and the idea of owning a bee farm in the country is a sustaining fantasy. I try not to discourage them, but sometimes I have to. Almost any kind of farming dooms newcomers to bankruptcy these days, particularly bee farming, because of the market conditions. About the only quicker way to go broke right now is to raise pigs, so when a man came to seek my advice a month ago and told me he was ready to spend ten years' worth of savings from a factory job to buy a farm where he would raise bees *and* pigs, I had to admit it was the worst idea I had heard in a long time. He went away saddened; I don't know whether he bought the farm or not.

The Ozarks, wild, undeveloped, inhospitable, keep being discovered. A lot of people who figured it was better to be poor in the country moved here during the 1930s; others, richer, thought FDR was the devil incarnate and wanted to put their wealth in land before he could take it away from them. Since then waves of people who find the cities too complicated have come here, meaning to lead lives of simplicity. What they have not yet discovered is that a life is as simple or as complicated as the person living it, and that people who have found life in the city over-whelming will find it even more so here, where it is much harder to make a living. When a person has money coming in regularly, his mistakes may make him unhappy but they do not threaten his survival. Here, where there is little money, every decision counts and there is no room for mistakes.

The people who live here have been idealized by the back-to-the-landers while they still lived in cities, but they are not simple people at all. Ozarkers lead lives as complicated as those of people anywhere else. However, they are competent and resourceful about living in these hills; they are quiet about it, too, so it looks easy and . . . simple.

Ozarkers are of mixed mind about the newcomers. The Simple Lifers always have a theory or two that they are not at all shy about expounding—theories which differ in details but always come down to knowing better how to live in the country than the peasants do. Understandably, Ozarkers often resent this. On the other hand, over the years the Simple Lifers have come to represent a cash crop like no other. They come with a bundle of savings to back up their theories, and it goes so quickly that it is good not to let one's resentment show too much. It is better to be able to profit from them before they return, dreams shat-tered, to their cities and paychecks.

Ozarkers have a saying about back-to-the-landers: The briars get their clothes, the hillbillies get their money and they leave with an empty suitcase in their hands.

I've lived both sorts of lives and have sympathy with both points of view, so I try to be tender with the Simple Lifers when they show up on my doorstep. But they take up time, sometimes too much of my time, and I have had to learn to say no and not feel guilty about it when I have to get on with my own not-so-simple life.

At the end of the day last summer a woman telephoned when I was in the middle of the honey harvest. It was late, but I was still out in the honey house and answered the telephone there. My feet ached from standing on the concrete floor all day, and I was sticky and tired. She introduced herself, said she had heard about me and wanted to ask a few questions. Should she tip the man who drove the road grader? No, she should not. It would embarrass him. Might she come out for a visit? No, she might not, I said as kindly as possible; I was just too busy. She sounded so lonely that I asked a few friendly questions; she told me she had just bought a farm and moved here, that she had lived all over the world, most recently in New York on the city's Upper West Side, and that life was very different here.

"And now you are experiencing culture shock?" I asked, trying to picture the move from a genteel old brownstone to an Ozark farm with an outhouse.

"Well, it's not as bad as Afghanistan," she said, trying to sound cheerful.

The woman really was lonely and overwhelmed. I tried to find some consoling words, and invited her to call later in the year when I might not be so busy. I was truly grateful for her call. For a few minutes I had forgotten my aching feet, and I have lived here so long that I sometimes forget how life in the Ozarks appears to outsiders. It was good to be reminded.

My brother Bil is a large man with unruly dark brown hair, and I am a small woman with unruly light brown hair. In addition to the generally rumpled appearances we present, we share a similar outlook on the world and enough affection toward and understanding of each other to be able, sometimes, to do things together without ever saying a word. Yesterday, while we were going down to the river to look at some ferns, we acted in instant and silent cooperation to save a newborn fawn.

We were walking on an old road bordered by a fence. Tazzie and her brother, Xenas, Bil's dog, were playing rowdy puppy with one another and did not immediately notice that Andy and Chocolate, Bil's older dog, were examining something bunched up in the tall grass next to the fence. But Bil and I did. We identified it—faster than the dogs did—as a recently born fawn, his absolute motionlessness his only defense. The dogs were on the brink of understanding that this dappled something was potential excitement. There is no way to know whether this understanding was complicated and classificatory—"Aha! Here we have the small specimen of that larger creature whom it is so delightful to chase"—or whether it was simply "Ah! What have we here? I do believe this small spotted object might be roused, run to the ground and torn to bits." Whatever the process, Bil and I knew that whichever conclusion the four dissimilar dogs arrived at, it would change them into a murderous pack. In the fraction of a second before this understanding was reached, Bil and I spoke sharply enough to our dogs to divert their attention to obedience, and I hustled them on down the road while Bil scooped up the fawn and unceremoniously stuffed him through the woven wire fence, the fawn's gangly, fragile legs getting in the way everywhere. On the other side of the fence he was safe from the dogs, and he ran off unsteadily.

He was the youngest fawn Bil or I had ever seen in the wild. He may have weighed ten pounds, but probably not more. His glossy, reddish-brown coat was freshly spotted with white. He was a newcomer to the world. We both would have liked to have

watched him longer but the doe must have been nearby and seen her offspring's rescue. Although she was invisible to us, she would not long be so to the dogs, and we wanted to hurry them away before they discovered her or recollected and understood the fawn, so we walked briskly on down the road to the limestone ledge where the ferns grow.

This peninsula is home to a number of natural rarities. A seldom-seen wild orchid grows here. Pileated woodpeckers, large prehistoric-looking birds with big red crests, are common here and uncommon elsewhere. And walking ferns, scarce in many places, grow here in thick, matted profusion.

The walking fern, *Camptosorus rhizophyllus,* is the only species on the American continent of this small genus. It is a curious little fern in other ways, too. All true ferns reproduce by spores, alternating asexual and sexual generations, but the walking fern has discovered a second means of reproduction, a short cut to proliferation. It is low growing, and when its long, narrow, fine, pointed, arched leaves touch the ground, new plants spring up from their ends, plants which in turn sprout more plants on their leaf tips. Old-established parent ferns are often surrounded by several generations of attached plants "walking" away from the center.

There was this sort of colony of walking ferns at the place where Bil and I were going. They grow under the trees near the river, in cool shade where underground water seeps, keeping the rocks moist. The trees and undergrowth screen them from the view of anyone floating by in a boat or canoe, but the evergreen, leathery-leafed ferns are not likely to attract notice anyway among the mosses and lichens that also grow on the rocks where the ferns find roothold.

Bil and I scrambled down the cliffs. Bil lit a cigarette and talked about ferns, while the dogs raced on ahead to the river, splashing about chasing each other and fish, real and imaginary.

Walking ferns, like other true ferns, bear sori, or fruit dots, on the underside of their leaves. The sori contain sporangia and the sporangia contain spores. When they are ripe and the surrounding air dry enough, the spores burst forth and disperse on air currents. A spore lucky enough to settle in a shady, moist spot puts down a rootlike hair to hold it in place, while cell after cell of green plant tissue grows from the spore. On this tiny gametophyte sprout separate male and female organs containing sperm and eggs. Moisture breaks open the mature male antheridium; at the same time the female archegonium opens and exudes a chemical attractive to the sperm, which swim toward the egg. From the fertilized egg a new fern grows.

In this manner ferns reproduce without seeds, clinging to a more ancient way of duplication worked out even before Devonian days when the first fossil fern records appear, and long before there were flowering plants with seeds.

In Europe in the Middle Ages the reproduction of flowering plants by seeds was understood, but thought to be universal for all plants. No one, of course, had ever seen the seed of a fern plant, but in those days, when magic could explain all, this simply meant that fern seed was invisible. As a result by a neat transfer of properties, it was held that fern seed could convey invisibility. It was a custom to spread linen cloth upon the ground on the eve of Midsummer Night to catch fern seed, for there are any number of occasions on which invisibility could have its advantages.

As late as Shakespeare's day, Gadshill, laying the groundwork for Hal's comic robbery in *Henry IV,* Part I, boasted, "We steal as in a castle, cocksure; we have the receipt of fern-seed, we walk invisible." To which the inn's chamberlain scornfully replies, "Nay, by my faith, I think you are more beholding to the night than to fern-seed for your walking invisible."

Our talk, Bil's and mine, turned to other plants, and as we walked down to the gravel bar on the river Bil began telling me about the trees of New Zealand. He was just back from there, having collected material for a magazine article, and was full of his unwritten story. Bil has a compelling personality. He is a good talker and even better storyteller, and while he talked and smoked the banks of the river disappeared and we were, in truth, in New Zealand. I sat and listened to him tell of exotic trees and of the people whom he had met who held them in sacred awe, and of others who found them of commercial interest. At last his stories were told, and we sat in silence. Bil smoked one more cigarette and we both watched a pair of kingfishers flying in Union-blue formation, quartering the river looking for minnows, their rattling call sounding as though it were a necessary mechanical part of their synchronized flight.

At last we whistled up our dogs and walked back up to the cabin. We stopped by the fence where the fawn had been, but there was no sign of the youngster or his mother. The bent grasses where he had lain had sprung back, and even his scent must have been gone, for the dogs waiting near us, quiet and tired, took no notice.

A year ago, on an afternoon late in springtime, I was walking on the dirt road that cuts across the field to the beehives. I noticed a light-colored, brownish dappled something-or-other stretched across the roadway ahead of me, and decided that it was a snakeskin. I often find them, crumpled husks shed by snakes as they grow. They are fragile and delicate, perfect but empty replicas of the snakes that once inhabited them. I started to turn it over with the toe of my boot, but stopped suddenly, toe in air, for the flecked, crumpled-looking empty snakeskin was moving.

It gave me quite a start and I was amused at my own reaction, remembering that Ronald Firbank wrote somewhere that the essence of evil was the ordinary become unnatural, the stone in the garden path that suddenly begins to move.

I squatted down to see what queer thing I had here, and found that my supposed snake skin was a mass of maggoty-like caterpillars, each one no more than half an inch long. They were hairless, with creamy white smooth skin, black heads and brown stripes along their backs. They were piled thickly in the center, with fewer caterpillars at the head and rear end of the line, which was perhaps eighteen inches long. They moved slowly, each caterpillar in smooth synchrony with its fellows, so that a wave of motion undulated down the entire length of the line.

They seemed so intensely social that I wondered what they would do on their own. I gently picked up half a dozen or so, and isolated them a few inches from the column. Their smooth, easy movements changed to frantic, rapid ones, and they wriggled along the ground quickly until they rejoined the group. They certainly were good followers. How did they ever decide where to go? The single caterpillar in the lead twisted the forepart of his body from side to side as though taking his bearings; he appeared to be the only one in the lot capable of going in a new direction, of making a decision to avoid a tuft of grass here, of turning there. Was he some special, super-caterpillar? I removed him from the lead position and put him off to the side, where he became as frantic as had the others, wriggling to rejoin the group somewhere in the middle, where he was soon lost to view, having turned into just another follower. At the head, the next caterpillar in line had simply assumed leadership duties and was bending his body from side to side, making the decision about the direction the column was to take. I removed three leaders in a row with the same result: each time, the next caterpillar in line made an instant switch from loyal and will-less follower to leader.

What were they doing? Were they looking for food? If so, what kind? What manner of creature were they? The beework that I had set out to do could wait no longer, so I went back to the hives. When I returned along the road, the caterpillars, if that is what they were, had disappeared.

Back in my cabin, none of the books on my shelves were much help explaining what I had seen, except one by Henri Fabre, the nineteenth-century French entomologist who had conducted one of his famous experiments with pine processionaries, one of the Thaumatopoeidae. Fabre's caterpillars were *Thaumatopoea proces-sionea,* "the wonder maker that parades"; eventually they become rather undistinguished-looking moths.

The pine processionaries are a European species, but their behavior was similar to that of my caterpillars, although not identical. Pine processionaries travel to feed in single file, not massed and bunched, but they do touch head to rear and have only one leader at a time. Fabre found them so sheeplike that he wondered what they would do if he could somehow manage to make them leaderless. In a brilliant experiment, he arranged them on the upper rim of a large vase a yard and a half in circumference, and waited until the head end of the procession joined the tail end, so that the entire group was without a leader. All were followers. For seven days, the caterpillars paraded around the rim of the vase in a circle. Their pace slowed after a while, for they were weary and had not been able to feed, but they continued to circle, each caterpillar unquestioningly taking his direction from the rear of the one in front, until they dropped from exhaustion. However, even Fabre never discovered what it was that could turn one caterpillar into a leader as soon as he was at the head of the line.

It was not until several months later, when I was talking to Asher, that I was able to find out anything about the caterpillars I had found in the roadway. He said that I probably had seen one species or another of sawfly larvae. They are gregarious, he told me, and some are whitish with brown stripes. Sure identification could only be made by counting the pairs of their prolegs, and of course I had not known enough to look at them that closely. Asher said that they were a rare sight and that I would probably never see them again, but if I did I should gather up a few and put them in a solution of 70 percent alcohol; then he would help me identify them. He had read about Fabre's experiment too, but knew nothing more about their behavior.

He added, "If you ever find out what makes processionary caterpillars prosesh, please enlighten me. Maybe it's the same thing that makes people drive in Sunday traffic or watch TV or vote Republican."

It is springtime again. I would like to count the caterpillars' prolegs and am prepared to pickle a few to satisfy my curiosity, but mostly I should just like to watch them again. This time I should let the beework go. I should like to know where these caterpillars go, and what it is they are looking for. I wonder if I could divide them up into several small columns that would move along independently side by side. I have more questions about them than when I first saw them.

This spring I often walk along, eyes to the ground, looking for them. There may have been nobler quests—white whales and Holy Grails—and although the Ahabs and Percivals of my acquaintance are some of my most entertaining friends, I am cut of other stuff and amuse myself in other ways. The search for what may or may not be sawfly larvae seems quite a good one this springtime.

About the Author

*Sue Hubbell was born in Kalamazoo,
Michigan. Before becoming a commercial
beekeeper in Missouri, she was a bookstore
manager and a librarian.*